The Aboveground Steel Storage Tank Handbook

The Aboveground Steel Storage Tank Handbook

BRIAN D. DiGRADO
and
GREGORY A. THORP, P.E.

Published by John Wiley & Sons, Inc., Hoboken, New Jersey
Published simultaneously in Canada

For general information on our other products and services, or technical support, please contact our Customer Care Department within the United States at 800-762-2974, outside the United States at 317-572-3993 or fax 317-572-4002.

Wiley also publishes its books in a variety of electronic formats. Some content that appears in print may not be available in electronic books.

For more information about Wiley products, visit our web site at www.wiley.com.

Library of Congress Cataloging-in-Publication Data:

10 9 8 7 6 5 4 3 2 1

To the two most important women in my life: my wife, Kelly DiGrado, who is always there for me, and my mother, Dorothy Meyerdierks, who has instilled in me the drive and determination to go the extra mile.

—Brian D. DiGrado

To my wife, Cindy Thorp, whose endless support and patience are greatly appreciated, and to my parents, Rand and Dianne Thorp, who have given me the motivation always to do my best.

—Gregory A. Thorp

Contents

Preface

Storage tanks have been in existence for thousands of years, storing all kinds of products that society has required to survive. Early wooden tanks used to contain water, alcohol, whale oil, and food were perfectly adequate in their day. The first successful oil well was drilled in Pennsylvania in 1859; with increasing industrialization, our society's dependence on oil has increased dramatically since then. Today, oil provides gasoline, jet fuel, heating oil, kerosene, petrochemicals, and many synthetic materials. Try to estimate how many items just in your home, office, and car are made of plastic, which is derived from oil. The point is that all these products must be stored during the distribution process, and steel aboveground storage tanks play a major role in such storage.

We have tried to compile pertinent and relevant information on both field-erected and shop-built steel aboveground storage tanks (ASTs), while looking at the state of the AST industry in the 1990s. We have developed a handbook that is easy to read and to comprehend, and one that describes the overall AST market.

The unique aspect of this book is the focus on field-erected and shop-built ASTs in a single source. During our extensive research effort, it was clear to us that all prior publications on the subject of ASTs focused on either field-erected ASTs or shop-built ASTs; rarely would both technologies be mentioned in the same context.

Although we discuss the vast majority of AST standards and regulations, we focus on the safe storage of petroleum and petrochemical products in ASTs for a number of reasons. Mainly, the various environmental regulations were written to prevent the discharge of petroleum and petrochemicals into the ground or waterways and to reduce emissions from volatile organic compounds (VOCs) from ASTs and other sources. Also, the majority of welded steel ASTs are installed directly for the petroleum industry, whereas most of the remaining ASTs are used in related areas such as chemical and hazardous waste industries. Finally, the standards and related AST products do not vary greatly between the different industries.

We found that the most logical method of presenting the information was to break the AST market into four manageable sections, on the AST market, regulations, standards, and products. The AST market section offers an overview of the AST industry through the use of statistical analysis and by exploring environmental impact issues. The regulations section highlights essential environmental and fire regulations that affect ASTs and explains the responsibility of the AST owner in meeting these requirements. The standards section discusses the national AST manufacturing and fabricating standards employed by AST fabricators in producing a finished AST product. The products section compares and contrasts the latest developments in the steel storage tank market and shows the potential AST owner which product is best suited for a particular application. Finally, in keeping with the handbook nature of this text, the appendixes cover practical and useful topics.

Every effort has been made to produce an accurate publication, but the aboveground storage tank market is diverse and complicated, with continuous updating of codes and standards. We are confident that, upon review of this book, you will have a solid understanding of the AST marketplace and be able to make a sound purchasing decision on your next, or first, steel aboveground storage tank.

Acknowledgments

It is a pleasure to acknowledge all the people who have assisted and supported our effort to develop the most comprehensive book ever written on the subject of aboveground storage tanks. Several people have read various chapters and provided positive and useful suggestions. In particular, thanks are due to Wayne Geyer, P.E., of the Steel Tank Institute; Lois Epstein, P.E., of the Environmental Defense Fund; and Kelly Karr, P.E., of Environment Canada. Additionally, our gratitude goes to Van Nostrand Reinhold for making this project a reality.

Also, we thank the following associations for allowing us to use their information in producing this text: the American Petroleum Institute, the Steel Tank Institute, the Petroleum Equipment Institute, the National Fire Protection Association, the International Fire Code Institute, and Underwriters Laboratories.

Finally, we are most grateful to Kelly DiGrado and Cindy Thorp. Over the course of this long project, their faithful support and their dedication to proofreading have been invaluable.

PART I

The AST Market

CHAPTER 1

The Aboveground Steel Storage Tank Market

To place the aboveground storage tank (AST) market in perspective, we must study statistics related to this industry. Numerous publications, industry groups, and the government have conflicting versions of the exact size and economic impact of this market. Consequently, we have attempted to offer an unbiased and fairly conservative view of the AST market from a variety of reputable sources.

ASTs have been storing petroleum products for nearly a century without much fanfare (see, for example, Photo 1-1), but environmental concerns and the increasing age of the existing storage tank population suggest that future applications of ASTs will be a chief focus of all AST owners and operators. The U.S. Department of Energy has estimated that over 700 million gallons of petroleum is consumed daily, an amount equal to over 255 billion gallons of petroleum annually.[1] The U.S. economy is strongly influenced by our dependence on petroleum products, and all petroleum products

PHOTO 1-1.
ASTs at a petroleum storage terminal.

spend some time in ASTs. Moreover the Environmental Protection Agency (EPA) states that over 50% of the U.S. population depends on groundwater as its main source of drinking water. So, as the ASTs continue to age, we must ask what percentage of existing tanks are leaking today and potentially damaging our groundwater supplies.

Shop-Built ASTs vs. Field-Erected ASTs

Over the years, numerous publications have printed articles and studies on the AST marketplace, covering topics ranging from regulations to manufacturing standards. Unfortunately, these publications commonly cover either shop-built ASTs or field-erected ASTs; only very infrequently will an article discuss both technologies. In this book, every effort has been made to cover both shop-built and field-erected AST issues independently, a unique approach in this industry.

For ease of definition, a shop-built AST is generally considered to be available in capacities less than 1000 barrels (42,000 gallons) and 15 feet in diameter, whereas a field-erected tank is usually available in capacities greater than 1500 barrels (63,000 gallons) and 16 feet in diameter. ASTs with capacities between 1000 and 1500 barrels can be either shop-built or field-erected, according to product applications and other external factors. (*Author's note:* The "barrel" measure goes back to the first discovery of oil in the United States in 1859. The only practical way to transport oil was in 42-gallon wooden whiskey jugs that were abundant at the time. Although the modern "barrel" holds 55 gallons, the 42-gallon traditional measure is still used throughout the world.)

Another difference between the technologies is the degree of engineering required in the products. Shop-built ASTs are generally preengineered, standard products that are usually purchased "as is," and the application must fit the predesigned parameters of the shop-built AST. Although shop-built ASTs can be custom-designed, the degree of engineering required is somewhat limited because of the small size of the storage vessel. Conversely, field-erected ASTs are always custom-designed and engineered, offering greater flexibility than shop-built ASTs in meeting the requirements of the application. Table 1-1 highlights some of the differences between the AST configurations.

Table 1-1. Shop-built ASTs vs. field-erected ASTs

Shop-Built ASTs	Field-Erected ASTs
Preengineered	Custom-engineered
Shipped as one unit for installation on site	Shipped as multiple parts for field erection
1000-barrel capacities and less	Greater than 1500-barrel capacities
Standard products/little flexibility	Custom-designed/great flexibility
Quick delivery/availability	Long lead times/crew mobilization

Number of ASTs

The American Petroleum Institute (API) in a mid-1989 study, which is the most current, concluded that there were approximately 700,000 ASTs in the United States storing petroleum products as listed in Table 1-2.[2] ASTs storing petroleum are located in approximately four areas: refineries, marketing terminals, transportation facilities, and production areas. These ASTs can be further divided into two groups: field-erected ASTs (over 1500-barrel capacities) and shop-built ASTs (under 1000-barrel capacities). Therefore, approximately 75,000 ASTs would be classified as field-erected, with the remaining 625,000 ASTs shop-built. (Photo 1-2 shows shop-built ASTs and an open-top floating roof tank.)

Table 1-2. Number of petroleum ASTs per API

Sector	Tank capacity (bbl)	AST quantity	Median diameter (ft)
Marketing	26 to 500	64,793	10
	500 to 1,000	4,417	15
	1,000 to 10,000	7,434	25
	10,000 to 100,000	11,469	70
	100,000 or more	416	220
	Total	**88,529**	
Refining	26 to 500	3,913	12
	500 to 1,000	2,460	15
	1,000 to 10,000	9,665	30
	10,000 to 100,000	11,629	90
	100,000 or more	2,060	220
	Total	**29,727**	
Transportation	26 to 500	694	10
	500 to 1,000	307	15
	1,000 to 10,000	1,468	35
	10,000 to 100,000	5,048	120
	100,000 or more	1,680	220
	Total	**9,197**	
Production	26 to 500	510,045	10
	500 to 1,000	37,628	16
	1,000 to 10,000	23,946	30
	10,000 to 100,000	974	70
	100,000 or more	27	200
	Total	**572,620**	
All Sectors	26 to 500	579,445	
	500 to 1,000	44,812	
	1,000 to 10,000	42,513	
	10,000 to 100,000	29,120	
	100,000 or more	4,183	
	Total	**700,073**	

Source: American Petroleum Institute.

It should be noted that the API studies assume all these tanks are vertical in orientation. Although almost all field-erected tanks would fall within this definition, shop-built ASTs are commonly horizontal in configuration. So, the question becomes: Are the API studies somewhat flawed, or are there many more horizontal shop-built ASTs that are unaccounted for? The answer to this question could significantly alter the overall number of ASTs in the United States.

In 1992 the Steel Tank Institute (STI) conducted a representative survey[3] of member companies that dealt with the types of shop-built

PHOTO 1-2.
Vertical shop-built ASTs and an open-top floating roof tank at a petroleum terminal.

ASTs they were manufacturing as outlined in Table 1-3. The STI survey stated that the member companies had seen an increase in AST production from 1991 to 1992 of almost 34%.

Table 1-3. 1992 STI member shop-built AST production survey

Percent of production	Design type	Configuration
65.28%	Single wall	horizontal or vertical
10.22%	Double wall	horizontal or vertical
23.64%	Integral steel dike	horizontal
0.85%	Fire-protected	horizontal

Source: Steel Tank Institute.

A Steel Tank Association (STA) member survey conducted from the fourth quarter of 1990 to the fourth quarter of 1991[4] suggests that on average 16% of the single wall and double wall designs (see Figure 1-1) or 12.1% of all the shop-built ASTs are vertical (i.e., 65.28% + 10.22% = 75.50%, multiplied by 0.16 to give 12.1%; refer to Table 1-3). Therefore, the remaining 87.9% of the shop-built

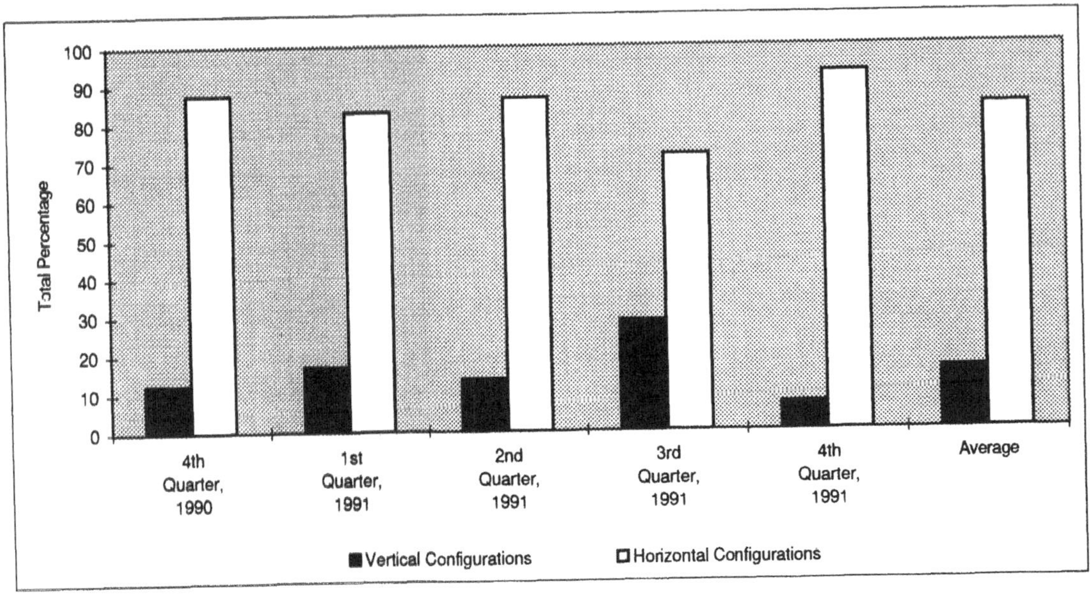

FIGURE 1-1.
Vertical vs. horizontal configurations, 1990/1991.

AST must be horizontal in configuration and would not have been accounted for in the API studies. With the increase in production of fire-protected ASTs beginning in 1993/1994 due to variance in fire codes, the low percentage of fire-protected tanks listed in the 1992 STI survey results would be much higher today and would increase the horizontal AST percentage even more.

Another source of AST numbers is the Independent Liquid Terminal Association (ILTA), which has a substantial number of ASTs among its member companies. ILTA is made up of independent terminals and storage facilities around the United States that lease their tanks to oil companies and others for storage. ILTA has 82 member companies operating in 45 states that have approximately 9000 ASTs in 360 facilities with total storage of 248 million barrels or 10.4 billion gallons of product.[5]

An industry that uses a substantial number of ASTs is the pulp and paper sector. According to industry data, there are 360 companies in the United States operating over 1300 pulp and paper mills.[6] Each facility has numerous ASTs of varying configurations in operation, including both shop-built and field-erected designs. A conservative estimate of the number of ASTs in use at these facilities would be in the 12,000-unit range.

The Petroleum Marketers Association of America (PMAA) conducts an annual survey of its 11,000-member organization.[7] In 1994, PMAA members operated 10,500 bulk storage facilities (bulk plants and terminals) with a total bulk storage capacity, consisting mainly of ASTs, of 1.23 billion gallons. The average member's bulk storage facility had five tanks with a 118,000-gallon aggregate capacity.

The Federal Aviation Association (FAA) estimated that 1992 general aviation fuel consumption in the United States was 807.8 million gallons, which consisted of both jet fuel and Avgas. As most airports and fixed-base operators (FBOs) utilize ASTs, a significant number of tanks are in operation.

Finally, the Society of Independent Gasoline Marketers of America (SIGMA) conducts an annual statistical report of its 250 members.[8] In 1993, SIGMA members sold 25.5 billion gallons of motor fuel. A total of 41% of the SIGMA member companies operate bulk plants and terminals that use ASTs.

After review of the various surveys and industry reports, and researching of pertinent industry papers, a more accurate total of ASTs is in the 950,000 to 1,100,000 range. Shop-built ASTs account for approximately 85% of this total (807,500–935,000), with field-erected ASTs accounting for the remaining 15% (142,500–165,000).

Age Distribution of ASTs

A major factor in determining the state of the AST industry is to investigate the average age of the AST population. The API study previously mentioned did a percentage breakdown of the four main industry sectors, as shown in Table 1-4. Conclusions can be drawn from these data, which show that in the marketing, refining, and transportation sectors an average of 90% of all ASTs are over 11 years of age. The production sector has many newer ASTs that lower the 11-year-old statistic to 68% of all ASTs. The U.S. Environmental Protection Agency (EPA) has stated in many proposed regulations appearing in the Federal Register that ASTs begin to deteriorate at 15 years of age. If this estimate is accurate, then numerous ASTs are currently in need of repair or retrofitting.

Table 1-4. AST age distribution, petroleum industry

Sector	0–10 years	11–20 years	21–30 years	31–40 years	41+ years	Unknown
Marketing	10.82	15.21	23.91	17.60	23.89	8.57
Refining	6.95	11.59	16.19	23.13	37.44	4.69
Transportation	9.52	14.96	19.91	22.79	28.99	3.83
Production	37.10	29.31	19.58	5.04	2.29	6.62
Total	32.13	26.59	19.99	7.63	6.86	6.79

Source: American Petroleum Institute.

Size of AST Market

An AST industry report by the consulting firm of Rauscher Pierce Refsnes, Inc. (RPR),[9] conducted in 1993, stated that the AST retrofit and inspection market is $1.1 billion annually with new AST construction at $400 million annually for an estimated total of $1.5 billion.

Although it is almost impossible to confirm, research suggests that the RPR estimates have a degree of validity and might be too conservative. Actually, the annual market for AST products and services is within the $1.5 to $2.0 billion range. This figure can be derived conservatively from the sales revenues of the largest AST product and service companies in the United States, and includes new construction, retrofit, repair, and inspection. Table 1-5 highlights the largest multi-facility AST fabricating and service companies. Obviously, there are numerous regional, single-location AST fabricating companies, in addition to those listed in Table 1-5, that offer good-quality services. In total, there are approximately 25 to 30 field-erected AST companies, and about 100 to 120 shop-built AST organizations in the United States. (Photo 1-3 shows both field-erected and shop-built ASTs at a marketing terminal.)

Table 1-5. Large multi-facility AST fabricating companies

Field-erected AST companies
CBI Industries, Inc. (Formerly: Chicago Bridge & Iron)
Pitt-Des Moines, Inc.
Astrotech International Corp. (Subsidiaries: Brown-Minneapolis Tank Company and HMT Inc.)
Matrix Service Company

Table 1-5. *(continued)*

Shop-built AST companies
Modern Welding Company, Inc.
Astrotech International Corp. (through subsidiary: Brown-Minneapolis Tank Company)
Highland Tank & Manufacturing, Inc.

PHOTO 1-3.
Field-erected and shop-built ASTs at a marketing terminal.

According to the *Fortune Magazine* industry report[10] (April, 1994), the single largest industry in the United States is petroleum refining, with sales of $398 billion and 546,000 employees. In 1993, the U.S. refining industry had 168 operating refineries (not including idle or shut-down refineries) producing 14.8 million barrels per day (b/d) of product while operating at 91% of capacity.[11] These refineries had working storage capacity in ASTs of 823.1 million barrels or 39.2 billion gallons of product. Table 1-6 gives a breakdown of storage capacity and number of refineries by state.

An annual study conducted by *Hydrocarbon Processing*[12] states that 1993 capital spending for new construction of ASTs in the United States was estimated at $125 million for the petroleum, petrochemical, refining, and gas processing markets, whereas world-

Table 1-6. AST capacities at refineries by state

District	State	Total storage capacity (barrels)*	Number of refineries (operating)
1	New Jersey	48,712,000	3
1	Pennsylvania	35,270,000	7
1	Other states[a]	28,610,000	5
2	Illinois	43,447,000	7
2	Indiana	24,833,000	4
2	Kansas	20,533,000	5
2	Ohio	15,920,000	4
2	Oklahoma	21,994,000	6
2	Other states[b]	33,613,000	9
3	Alabama	5,721,000	2
3	Arkansas	4,501,000	3
3	Louisiana	103,928,000	20
3	Mississippi	21,098,000	6
3	New Mexico	3,846,000	3
3	Texas	232,473,000	29
4	Colorado	4,462,000	2
4	Montana	8,523,000	4
4	Utah	6,554,000	6
4	Wyoming	7,281,000	4
5	Alaska	4,051,000	6
5	California	116,442,000	23
5	Washington	21,993,000	6
5	Other states[c]	9,251,000	4
United States totals		823,056,000	168

Source: Energy Information Administration, *Petroleum Supply Annual 1992*, May 1993.
*Total storage capacity of petroleum and petrochemical products.

(a) Other District 1 States: Connecticut, Delaware, Florida, Georgia, Maryland, Massachusetts, New York, North Carolina, Virginia, West Virginia.

(b) Other District 2 States: Kentucky, Michigan, Minnesota, Missouri, North Dakota, Tennessee, Wisconsin.

(c) Other District 5 States: Arizona, Hawaii, Nevada, Oregon.

wide capital spending for new construction of ASTs was in the $710 million range. On average, U.S. new tank construction since 1986 has approached approximately $182 million annually for these markets (HPI Market Data, 1986–1993).

In 1994, *Oil & Gas Journal* reported in a capital spending outlook special section that spending would be increased slightly from 1993 levels.[13] In the refining sector, 1994 capital spending was planned to be $5.4 billion, which was above the 1993 level of $5.36 billion. In the petrochemical sector, 1994 capital spending was expected to be $2.5 billion, which represented an increase from the $2.28 billion spent in 1993. Finally, in the marketing sector, 1994 capital spending was planned at $2.28 billion, showing an increase from the 1993 rate of $2.27 billion. Furthermore, this report stated that much of the spending would be in the area of environmental projects related to regulations such as the Clean Air Act. These environmental projects would include applications for ASTs, and showed that petroleum industry capital expenditures had remained fairly consistent during the first half of the 1990s.

A 1994 study conducted by Future Technology Surveys, Inc.[14] on aboveground storage tanks surveyed 22 AST fabricating company executives on the AST market. These respondents were asked to forecast the annual dollar value for new construction and fabrication of ASTs in 1993 and 1998. The median response was $150 million for 1993, with the amount increasing to $250 million by 1998. This estimate would equate to approximately a 14% annual growth rate for the five-year period. Also, the respondents estimated that the 1993 market for new ASTs was 13,000 units, compared to the 1998 market forecast of 25,000 units. Another question asked respondents to estimate the number of AST vendors in the United States in 1993 and 1998. The median response was 125 vendors in 1993, decreasing to 100 vendors in 1998.

Shop-Built AST Statistics

Although field-erected ASTs receive more national attention than shop-built ASTs due to a larger storage capacity and environmental disaster potential, shop-built ASTs (50,000 gallons and less) make up approximately 85% of the entire AST marketplace. (Photo 1-4, for example, shows a vertical shop-built AST installation.)

A study conducted for the Steel Tank Institute (STI),[15] published in 1993, of potential tank-purchasing decision makers offered some interesting conclusions about the future of the shop-built AST market. The survey showed a strong trend toward aboveground storage tanks from underground storage tanks (USTs) when respondents compared 1993 statistics with 1988 statistics. Of these, 48% stated that they had more ASTs in 1993, whereas only 15% said that

PHOTO 1-4.
Vertical shop-built AST installation.

they had more USTs. Also, 84% of respondents owned an AST, as compared to 60% of respondents who owned a UST. Furthermore, 50% of these respondents would purchase an AST next, whereas 20% would purchase a UST next.

The survey also revealed other important facts about the shop-built AST market, including the use of secondary containment and information about recent and future AST installations based on product and storage capacity. The findings in regard to the use of secondary containment systems were noteworthy. While 68% of respondents use some form of secondary containment on their shop-built ASTs, 31% of these systems have integral or attached secondary containment (i.e., double wall ASTs). Ironically, 29% of those surveyed do not have any type of secondary containment for their ASTs. The main products stored in recently installed ASTs are listed in Table 1-7(A), with the respondents' estimates of future product storage applications for shop built ASTs in Table 1-7(B).

To determine AST capacity trends, respondents were asked to list the most frequently requested AST capacities. The average respondent had nine ASTs in use. Table 1-8(A) highlights recently purchased AST capacities, and Table 1-8(B) lists respondents' anticipated future AST capacity requirements.

Table 1-7(A). Product storage, recent AST installations

Product	Percentage
Motor fuels	49%
Oil	17%
Chemicals	14%
Water	13%
Jet fuel	6%
Other	1%

Source: Derived from Steel Tank Institute Survey.

Table 1-7(B). Product storage, future AST installations (most common response)

Product	Percentage
Motor fuels	40%
Water	23%
Jet fuel	8%

Source: Derived from Steel Tank Institute Survey.

Table 1-8(A). AST capacity, recent purchases (most common responses)

AST capacity (gallons)	Percentage
Nominal 10,000	15%
Over 50,000*	15%
501–1,599	13%

Source: Derived from Steel Tank Institute Survey.
* Typically, ASTs over 50,000 gallons are field-erected.

Table 1-8(B). AST capacity, future purchases (most common responses)

AST capacity (gallons)	Percentage
Nominal 10,000	21%
Over 50,000*	14%
1,600–3,999	11%
6,000–9,999	11%

Source: Derived from Steel Tank Institute Survey.
* Typically, ASTs over 50,000 gallons are field-erected.

The *Journal of Petroleum Marketing* conducted a 1994 Buyers' Intention Survey of the PMAA membership.[16] Of interest were the results of the members' storage tank purchasing plans. A total of 80% of members planned to purchase storage tanks in 1994 according to Table 1-9. Also, a total of 45% of PMAA member environmental expenditures in 1994 would be directed toward replacing or increasing storage tank capacity. One limitation of these results is that the report did not state how many of the tanks would be ASTs.

Table 1-9. 1994 Storage tank purchasing plans

Number of tanks planned	Percent of members
None	20%
1–5	40%
6–10	24%
11–20	10%
>20	6%

Source: *Journal of Petroleum Marketing*, 1994.

Other AST Market Costs

The American Petroleum Institute, in a 1992 research study on "*Estimated Costs and Benefits of Retrofitting ASTs with Release Prevention Barriers*,"[17] offered additional cost statistics for the AST marketplace. A release prevention barrier (RPB) refers to materials and techniques used for preventing the release of product through the bottom of an AST or detecting the release through methods of leak detection. If the RPB is included in the design of a new AST, the increased cost is minimal. However, retrofitting of existing ASTs can be very expensive.

API member companies reported that retrofitting a small tank, 10 feet in diameter, would cost approximately $13,000, whereas retrofitting a large tank, 250 feet in diameter, can cost $1 million or more. Based on API's figure of 700,000 ASTs, this study estimates industry costs of $13 to $27 billion for retrofitting existing ASTs. API estimates that releases from petroleum ASTs due to equipment failure account for 3.7 million gallons per year, which could be substantially curbed with the addition of RPBs.

In a 1993 study by API,[18] statistics show that regulations will have a great effect on the AST marketplace. API estimates that the petroleum industry currently spends more than $8 billion per year on activities related to the environment, as indicated in Table 1-10, a historical record of expenditures.

Table 1-10. Petroleum industry expenditures, historical (dollars)

Petroleum industry	1973	1981	1990
U.S. exploration for oil	14.2 billion	45.0 billion	7.6 billion
Environmental activities	3.7 billion	8.4 billion	8.1 billion

Source: American Petroleum Institute.

In the same study, API estimates that environmental costs for the petroleum industry will range between $17 and $25 billion by the end of the 1990s, which would equate to approximately 14 to 17% of national environmental expenditures by the year 2000. New and potential regulatory costs are the driving force behind these anticipated environmental expenditures. To put these costs in terms of ASTs, Table 1-11 highlights certain regulations (both current and proposed) that have an effect on tank operators.

Table 1-11. Environmental compliance cost estimates—regulations

Requirement	Initial costs (millions)	Annual cost (millions)	Cost estimate source
SPCC program/ AST secondary containment	$579	$73	EPA
Oil spill response plans/ AST facilities	$107	$42	EPA
Clean Air Act Amendments	$3,142–$5,671	$214–$700	API/Bechtel
RCRA reauthorization	$3,377–$56,447	$117–$3,302	DOE/API
Clean Water Act reauthorization	$3,178	$32–$138	Bechtel
Proposed AST legislation	$9,159–$12,634	$270–$440	API

Source: Derived from API[19] and EPA.

Notes

1. National Energy Strategy: Powerful Ideas for America, First Edition 1991/1992, U.S. Department of Energy, DOE/S-0082P, February 1991, p. 74.
2. Aboveground Storage Tank Survey, prepared under contract by Entropy Limited (Lincoln, MA) for the American Petroleum

Institute (Washington, DC), API Publication number 301, April 1989, p. 1.
3. STI Member 1992 AST Production Survey, Steel Tank Institute, Lake Zurich, IL, 1992.
4. STA Market Survey of 4th Quarter 1990–4th Quarter 1991, Steel Tank Association, Lake Zurich, IL, 1991.
5. Bulk Liquid Terminals and Storage Facilities 1993 Directory, Independent Liquid Terminals Association, Washington, DC, 1993.
6. 1992 Lockwood-Post's Directory, published by Miller Freeman Publications, San Francisco, CA, 1992, pp. 5 and 11.
7. Petroleum Marketers Association of America 1993 Marketer Profile Survey, November 1993. Printed in *Journal of Petroleum Marketing,* BMT Publications, New York, NY, April 1994, p. 20.
8. Society of Independent Gasoline Marketers of America 1994 Statistical Report, Reston, VA, 1994, p. 2.
9. The Aboveground Storage Tank Report, Rauscher Pierce Refsnes, Inc., Dallas, TX, January 1993.
10. The Fortune 500 Industry Overviews, *Fortune Magazine,* New York, NY, April 18, 1994, p. 280.
11. U.S. Refineries & Capacities by State, NPN Market Facts 1994, *National Petroleum News Magazine*, Chicago, IL, June 1994, p. 130.
12. HPI Market Data 1993, a special report by *Hydrocarbon Processing*, Gulf Publishing Company, Houston, TX, 1993.
13. Capital Spending Outlook, *Oil & Gas Journal*, Pennwell Publications, Tulsa, OK, February 21, 1994, pp. 23–25.
14. Survey on Aboveground Storage Tanks, Survey Report #215, Future Technology Surveys, Inc., Lilburn, GA, December 1993, pp. 19 and 21.
15. Aboveground Storage Tank Boom—Will the Trend Continue?, prepared by the Steel Tank Institute, Lake Zurich, IL, January, 1993. Research for report conducted by Product Evaluation, Inc.
16. The *Journal of Petroleum Marketing* 1994 Buyers' Intention Survey of PMAA members, BMT Publications, New York, NY, March 1994, p. 43.
17. Estimated Costs and Benefits of Retrofitting Aboveground Petroleum Industry Storage Tanks with Release Prevention Barriers, American Petroleum Institute (Washington, DC), API Research Study #065, September 1992.

18. Costs to the Petroleum Industry of Major New and Future Federal Government Environmental Requirements, American Petroleum Institute (Washington, DC), API Discussion Paper #070R, October 1993.
19. Ibid.

CHAPTER 2

Environmental Issues for Aboveground Storage Tanks

Over the last decade, environmental concerns about the U.S. storage tank population have become a major issue for both environmentalists and regulators (see Photo 2-1, for example). In 1988, underground storage tank (UST) regulations were promulgated by the U.S. Environmental Protection Agency (EPA), that were intended to alleviate the threat of massive contamination due to catastrophic failure of aging bare-steel USTs. At that point, tank operators explored alternative ways to avoid costly UST rules where appropriate. This movement led to an increase in the production and unique design of ASTs. Today, with the regulated UST market under control, attention is being directed toward ASTs. The main question is: how great is the potential for environmental disaster?

In a 1993 study, the Environmental Defense Fund (EDF) reported that 20 to 25% of all ASTs and their associated piping were leaking.[1] Unfortunately, before we accept this statistic, it is impor-

PHOTO 2-1.
Many AST installations border waterways, which require proper environmental protection.

tant to see how it was derived. The EDF report based the leak rate on EPA estimates covering USTs for the 1988 regulations. The EPA's experts determined that the existing underground storage tank population—consisting mainly of bare steel tanks—exposed to soil would experience a leak failure rate of approximately 20 to 25%. EDF justified this number by stating that ASTs in contact with soil should experience the same failure rate. The problem is that EDF made the same erroneous assumption that API made (see Chapter 1) that all ASTs are vertical in orientation and, consequently, can be in contact with soil. To review, 85% of all ASTs are shop-built with about 12% of those being vertical configurations (horizontal shop-built ASTs are installed on some form of supports and would not come in contact with soil). The remaining 15% of ASTs are vertical field-erected tanks and are typically in contact with soil. Therefore, 12% of the shop-built ASTs and all of the field-erected ASTs (15%) would make a grand total of 27% vertical ASTs. According to EDF's estimate of a 20 to 25% leak rate, practically all of the vertical ASTs in the United States would be failing. This is simply not true.

Even though we find it necessary to discredit the EDF claim, environmental leak damage potential is a serious matter. As mentioned in Chapter 1, the American Petroleum Institute in a 1992 study estimated that releases from petroleum ASTs due to equipment failure account for 3.7 million gallons per year.[2] A 1994 American

Petroleum Institute survey report of 299 members revealed some startling statistics on groundwater contamination.[3] The survey concluded that 85% of U.S. refineries have groundwater contamination caused by AST systems. In addition, groundwater contamination is reported at 68% of marketing terminals and 10% of pipeline bulk storage facilities. An important distinction for ASTs is that leaking piping, not leaking tank bottoms, was considered the major cause of this contamination.

Let us turn our attention to some facts that can be confirmed. The above-referenced EDF report researched documented petroleum leaks from AST facilities, as reported to the federal Emergency Response Notification System (ERNS)[4] (see Table 2-1).

Table 2-1. Petroleum releases, 1987–1992

Category	1992	1991	1990	1989	1988/1987
Number of releases over 100,000 gallons	10	11	12	17	15
Number of releases between 10,000 and 100,000 gallons	59	N/A	N/A	N/A	N/A

Source: Emergency Response Notification Search and EDF report.

The figures given in Table 2-1 for 1992 could be expanded to show that the 10 large releases accounted for a total of 5 million gallons, whereas the 59 small releases accounted for 1 million gallons. Therefore, releases from AST facilities in 1992 officially reported to the ERNS accounted for 6 million gallons. From these figures it can be projected that from 1987 to 1992 approximately 6 to 8 million gallons per year of petroleum were released from AST facilities.

In a 1994 conference focusing on ASTs at the University of Wisconsin–Madison, industry leak detection experts claimed that the number of leaking ASTs that they inspect is less than 5% of all ASTs.[5] This is significant because the majority of the tanks that leak detection/testing companies inspect are suspected to be leaking. Unfortunately, there was no information given on the average size of the tested ASTs. So, should we believe that this 5% leak rate is for tanks with an average capacity of 2,500 gallons or for those of 1,000,000 gallons? Either way, most people would agree that it is important to strive for a leak-free AST community.

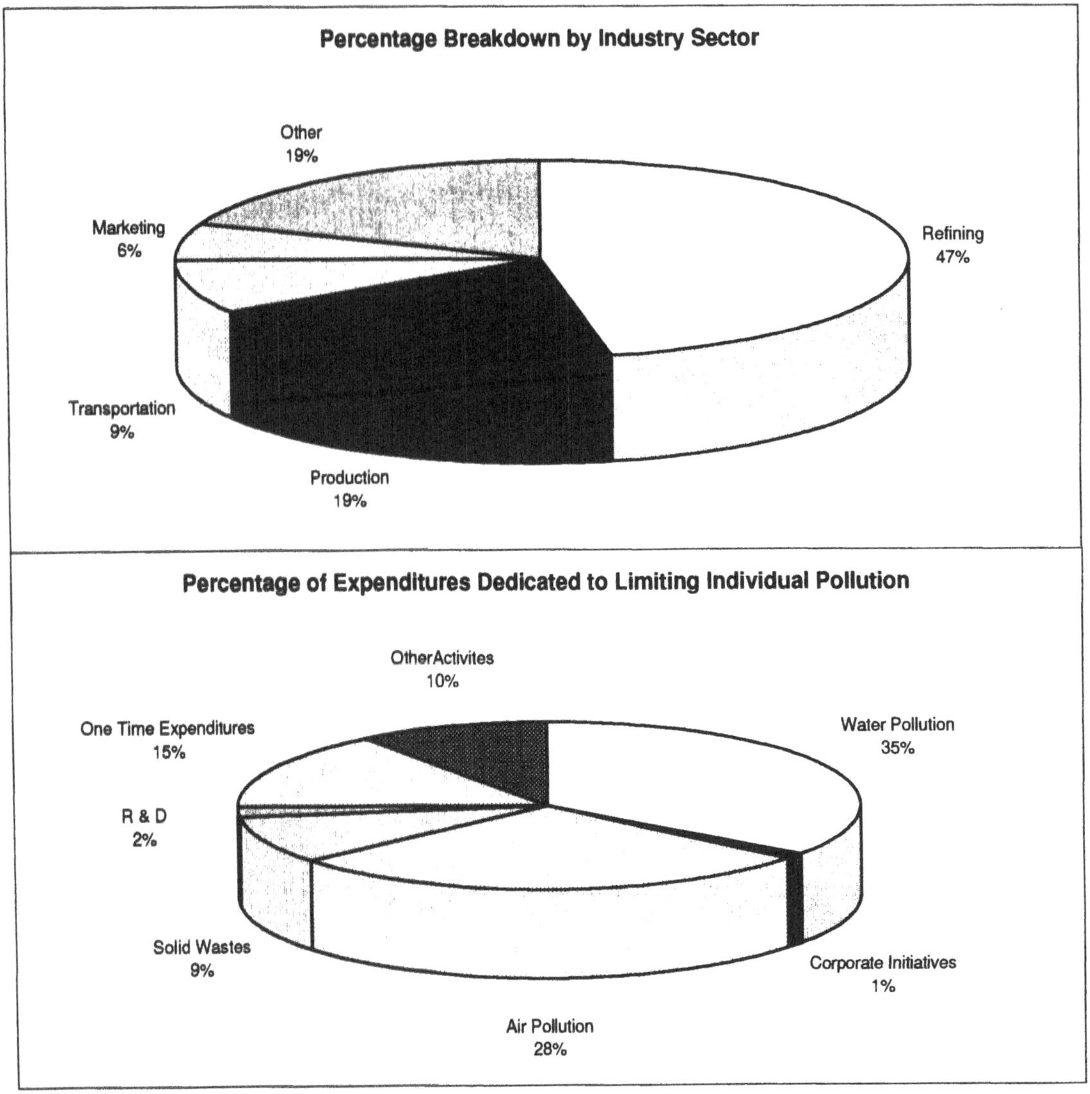

FIGURE 2-1.
1990 environmental expenditures, U.S. petroleum industry. (Source: American Petroleum Institute)

Much is being done by the petroleum industry to protect the environment from pollution. In 1990, the U.S. petroleum industry spent $7.8 billion on the environment (as illustrated in Figure 2-1), with 63% of the expenditures directly related to water and air pollution activities. During the same year, the entire budget of the EPA was only $5.8 billion.[6] Also, in 1990, U.S. petroleum refiners

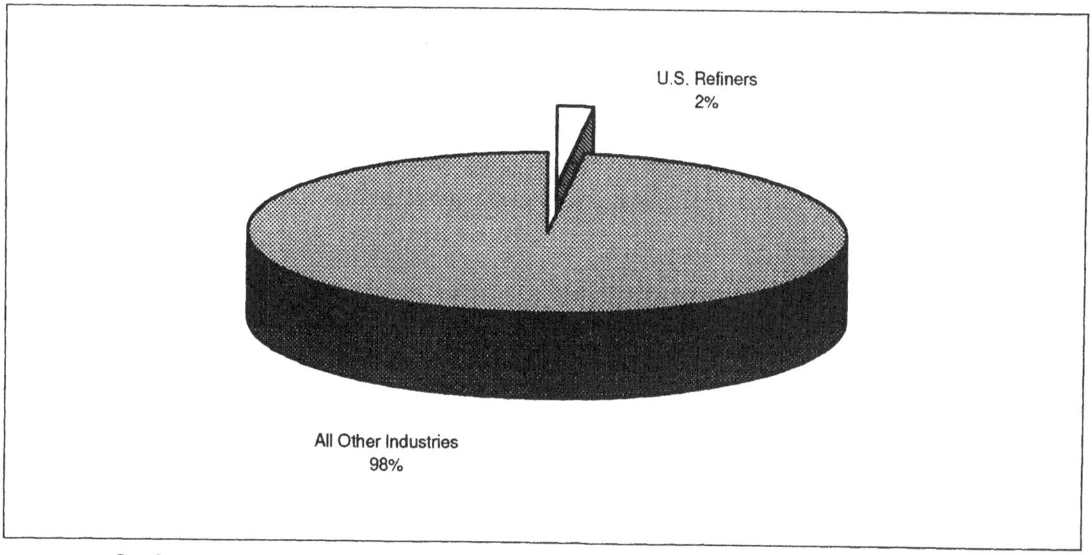

FIGURE 2-2.
1990 Chemical releases. (Source: American Petroleum Institute)

contributed only 2% of the total releases identified by the EPA Toxic Release Inventory, as shown in Figure 2-2.[7]

Environmental Disasters Involving ASTs

In 1992, Star Enterprise, a Texaco affiliate, settled with neighbors of the facility in Fairfax, Virginia for $200 million in compensation for decreased property values, medical expenses, and other damage claims. It was determined that petroleum found in a local creek, local sewers, and storm drains could be traced back to the Star terminal and might have been caused by leaking tank bottoms and associated piping.[8] In 1988, an Ashland Oil Company 40-year-old AST in Pennsylvania experienced a catastrophic failure, spilling 750,000 gallons of diesel fuel into the Monogahela River, threatening the water supply of about 1 million people in Pennsylvania, Ohio, and West Virginia.[9] Also in 1988, a Shell Oil Company facility in California had an AST failure, releasing 400,000 gallons of crude oil into a storm drain and damaging 150 acres of protected wetlands and 11 miles of shoreline.[10]

These environmental disasters affected a great number of people, and they received a lot of attention because of the size of the spills. Table 2-2 lists some lesser-known AST disasters of over 1 million gallons, which the Environmental Defense Fund has collected information about from the general press.[11] *Note:* neither the EPA nor any other federal agency keeps track of this information.

Table 2-2. AST releases over 1 million gallons

Facility	Location	Release size (gallons)
Chevron Refinery	El Segundo, CA	84–252 million
Tosco Refinery	Martinez, CA	28 million
Sun and Sinclair Refineries	Tulsa, OK	25–38 million
Amoco Refinery	Whiting, IN	17 million
Mobil Tank Farm	Brooklyn, NY	13–17 million
Southern Pacific Railroad Tank Farm	Sparks, NV	4–40 million
Clark (formerly Apex) Refinery	Hartford, IL	4 million

Source: Environmental Defense Fund, 1992.

Also, the Environmental Defense Fund has documented underground petroleum pools at facilities with multiple ASTs.[12] Table 2-3 shows the facility type and location of these petroleum pools that are larger than 100,000 gallons.

Table 2-3. Documented underground petroleum pools (larger than 100,000 gallons)

Location	Facility Type
Anchorage, AL	Air Force Tank Farm
Torrance, CA	Refinery
Port Everglades, FL	Transportation/Marketing
Granger, IN	Marketing
Cattlettsburg, KY	Refinery
Paulsboro, NJ	Marketing
Syracuse, NY	Marketing
Greensboro, NC	Marketing
Ponca City, OK	Refinery
Philadelphia, PA	Refinery
Spartanburg, SC	Marketing
Austin, TX	Marketing
Fairfax, VA	Marketing
Tacoma, WA	Marketing

Source: Environmental Defense Fund, 1993.

Fires Involving ASTs

On May 27, 1994, at a Shell Chemical plant in Belpre, Ohio, a processing unit exploded, causing five aboveground storage tanks to ignite at the facility's tank farm 600 feet away.[13] The five ASTs had a combined capacity of 4.5 million gallons, and were used to store the chemical styrene. Also, the EPA reported that almost one million pounds of other chemicals were released into the nearby Ohio River. Over 350 firefighters were needed to extinguish the fire. Three workers were killed during the explosion and resulting fires.

On January 2, 1993, at the Evergreen Terminal in Jacksonville, Florida on the St. John's River, a 100-foot-diameter internal steel pan floating roof AST, containing 2.3 million gallons of unleaded fuel, caught fire and began to overflow. The subsequent fuel spillage developed into a series of ground fires that spread beneath the bulk facilities piping and manifold assemblies. It took seven days to extinguish the fire and three more days to purge the AST. The cause of this disaster was overfilling the AST by 50,000 gallons and not opening the proper tank valves during the tank filling operation. The resulting explosion and fire claimed the life of the tank filling operator.[14]

On November 25, 1990, at the tank farm supporting Denver's Stapelton International Airport, a fire occurred that affected eight of the twelve ASTs on-site. The eight tanks contained jet fuel totaling almost 4.2 million gallons. The fire lasted 55 hours and destroyed or damaged seven ASTs while consuming 1.6 million gallons of jet fuel. This fire was linked to a damaged pump in a valve pit near the ASTs, which might have led to the initial leak and fire ignition. At the same time, fire investigators believe that a faulty pipe cracked, spilling fuel into the pit area. No fatalities or injuries resulted from this fire.[15]

In addition, the National Fire Protection Association produced a tank farm fire incident report in 1991 that detailed various disasters.[16] Some of the significant occurrences were as follows:

- In Arizona a worker was lowered into a 10,000-gallon vertical cone roof tank storing chemicals without respiratory equipment, and was subsequently overcome by fumes and lack of oxygen. Fire department personnel responded to rescue the worker. They decided to cut into the tank with a steel-bladed power saw to reach the worker more rapidly. After a few successful cuts, some sparks started a fire that led to a massive explosion. Fourteen firemen were injured, and one was killed.

- In New Jersey a tank farm with three field-erected tanks capable of storing between 1.7 million and 10.9 million gallons of gasoline each were located together in a common dike area. During the filling operation, one tank was not properly monitored and overflowed. Ignition from a nearby incinerator plant caused a major explosion that ignited the vapors of overflowing gasoline from the tank. The resulting explosion was responsible for damage up to 1500 feet from the source. Twenty-four injuries and one death were reported.
- During a fuel transfer operation, 15,000 gallons of gasoline overflowed onto the ground within a contained dike area. The operator was not monitoring the transfer, and the high liquid level alarm was not functioning. A telephone bell device was determined to be the source of the resulting explosion felt a mile away.
- At a bulk plant, a tank truck caught fire. The truck fire spread to engulf seven 20,000-gallon ASTs on unprotected steel supports, causing the supports to fail and spilling the entire contents of all seven ASTs.
- A vacuum truck pumping oil from a tank emitted fumes that came in contact with a water heater ignition spark 150 feet away. The surrounding dike was inadequate to contain the tank contents, with 100,000 gallons overflowing to the surrounding grass and brush area. The driver of the truck died.

Photo 2-2 shows a shop-built AST engulfed in flames.

Environmental Liability

The term "environmental liability" is an overused expression intended to instill fear in the hearts of AST owners and operators. Although that idea may be a little farfetched, the potential liability of a product release is a very serious matter to an AST owner.

As previously stated, Star Enterprise settled a lawsuit for $200 million in damages that did not include the retrofit costs to ASTs at its terminal. One wonders how many petroleum or chemical companies could afford this type of liability at their locations. What about the environmental liability of the AST facilities sitting on huge underground pools of oil (see Table 2-3)?

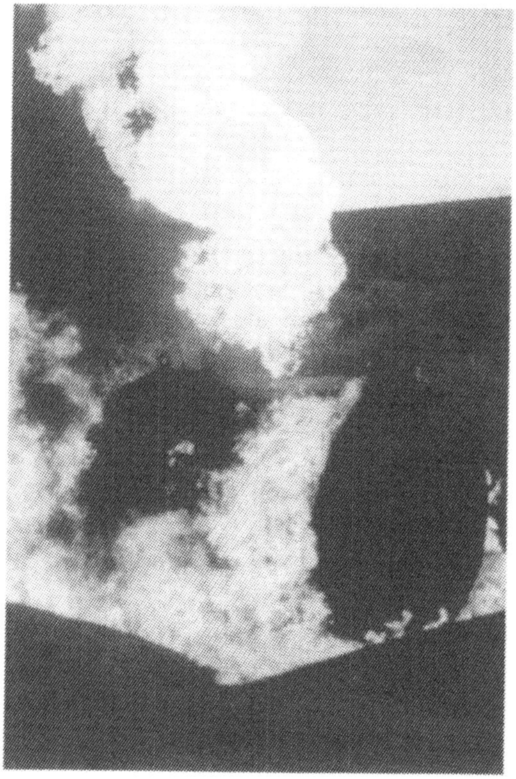

PHOTO 2-2.
Fires and ASTs are a dangerous combination. (Source: Steel Tank Association)

Federal and state regulators have attempted to establish funds that will finance product spill cleanups, but individual facilities still bear a large liability for releases. According to the Oil Pollution Act, the limit of liability for onshore facilities is between $8 million and $350 million, based on site-specific factors. If a discharge of product is directly caused by gross negligence, willful misconduct, or a violation of regulations, then the offending AST facility has unlimited liability for the cleanup. In addition, the Oil Pollution Act can administer civil penalties for product discharge or violation of regulations. These penalties may not exceed $10,000 per violation, with a maximum fine of between $25,000 and $125,000. If the penalty is for a product discharge, the limits are raised up to $25,000 per day of violation or $1,000 per barrel of oil (42 gallons). For failure to comply with regulations, a facility is subject to a penalty of up to $25,000 per day of violation. If the facility shows gross negligence or willful misconduct, the penalty will be not less than $100,000 and not more than $3,000 per barrel of oil.

This is just a sample of potential environmental liability under the Oil Pollution Act, but certain states also have individual requirements that have to be met.

The most important point about environmental liability is that most AST disasters are directly related to operator error or equipment failure. In the 1993 Steel Tank Institute survey of potential tank-purchasing decision makers discussed in Chapter 1, the subject of AST releases was addressed. Survey respondents reported that the main causes of a release were vandalism and overfilling of the tank, and 47% of ASTs that were damaged or had a release were field-erected.

By establishing proper worker training and safety programs (e.g., OSHA), operator error can be reduced. Appropriate security measures, including fences and adequate lighting, can curb the incidence of vandalism. Also, a regular AST maintenance program can detect equipment failure before it occurs.

Another way to offset a potential discharge is to incorporate some of the environmentally sound technology highlighted in the following section. These simple steps are the most effective methods of decreasing an AST facility's environmental liability.

Environmentally Sound Technology

An AST operator's goal is to safely and economically manage the tanks at its facilities while reducing the chance for any type of product release. The proactive way to accomplish this task is to utilize environmentally sound technology in the ASTs. This would entail installing release containment and emission containment systems when retrofitting an existing AST or when constructing a new AST. Many systems can operate effectively to reduce environmental liability, including double bottoms, floating roofs, alarms and valves, and fire suppression equipment. (See Photo 2-3.)

A principal cause of AST leaks is loss of integrity of the existing tank bottom. For older tanks, retrofitting the existing tank bottom with a new double bottom system with built-in monitoring capabilities could maintain the AST for additional future use and avoid the expense of constructing a new AST. In building a new AST, the additional cost of a double bottom system is insignificant at the time

PHOTO 2-3.
AST installation utilizing environmentally sound technology

of construction, and it will contain the product during a release without damaging the environment.

For emissions release problems, a floating roof with appropriate seals will significantly reduce or contain any unwanted vapors so that they are not discharged into the environment. Floating roofs can be retrofitted into existing ASTs or constructed as part of a new AST with a minimal cost difference.

As operator error is responsible for many AST releases, a few precautions will help to reduce disaster potential. During the filling or the product transfer operation, the AST can be equipped with a high liquid level alarm that will alert the operator when the tank level reaches 90% of capacity and an automatic shutoff valve that will close when the tank level reaches 95% of capacity. Both of these safety items, when maintained and functioning properly, will help to curb accidental operator error.

In case of fire, ASTs can be equipped with proper fire suppression equipment to help extinguish the fire. Firefighting foam discharge equipment, attached to the top of an AST and piped to a ground level water supply, are very effective in the event of a tank fire. The use of foam dams around the top of field-erected floating roof tanks will concentrate foam around the outer ring or seal area of the AST. Also, having a fire water tank on-site at a terminal would guarantee a continuous source of water for firefighting personnel. For shop-built ASTs, using a two-hour fire rated tank system can prevent catastrophic failure of the AST under fire conditions.

One often overlooked area of potential product releases is faulty pumps, valves, piping, and piping manifold systems. As in the Denver airport tank fire, a faulty pump and failed piping were determined to be the cause of that catastrophe. The use of properly installed and maintained piping manifold systems, including cathodic protection for underground steel piping, can be a positive step in avoiding releases. Pumps and valves should be a part of a facility maintenance plan and checked according to the manufacturer's recommendations.

As can be seen, environmentally sound technology is available that will reduce an AST operator's liability. These safety and release prevention measures can be retrofitted on existing ASTs or installed during the new tank construction process. The following chapters will explain various products and applications in detail.

Notes

1. LAST, But Not Least: Threats, Costs, & Answers, by Lois Epstein for the Environmental Defense Fund, Washington, DC, February 1993, p. 3.
2. Estimated Costs and Benefits of Retrofitting Aboveground Petroleum Industry Storage Tanks with Release Prevention Barriers, American Petroleum Institute, Washington, DC, API research study #065, September 1992.
3. API member survey of AST facilities, by SRI International, for American Petroleum Institute, Washington, DC, July 1994.
4. LAST, But Not Least: Threats, Costs, & Answers, by Lois Epstein for the Environmental Defense Fund (Emergency Response Notification System search performed by Larry O'Connor of ABB Environmental, January 1993), Washington, DC, February, 1993, p. 5.
5. Aboveground Storage Tank Conference, held at the University of Wisconsin, Madison, WI, March 9–11, 1994, cosponsored by the Steel Tank Institute, Lake Zurich, IL. Industry leak detection experts represented: Tracer Research Corp. (Tucson, AZ), Physical Acoustics Corp. (Princeton, NJ), and Leak Detection Systems, Inc. (Norwell, MA).

6. Assessing Our Environmental Performance Booklet, from Strategies for Today's Environmental Partnership Program, American Petroleum Institute, Washington, DC, 1991–1992, p. 8.
7. Ibid., p. 3.
8. LAST, But Not Least: Threats, Costs, & Answers, by Lois Epstein for the Environmental Defense Fund, Washington, DC, February, 1993, p. 8 and Tulsa Letter, by Robert N. Renkes, Petroleum Equipment Institute, Tulsa, OK, Volume 43, Number 20, September 22, 1993, p. 1.
9. Tulsa Letter, by Robert N. Renkes, Petroleum Equipment Institute, Tulsa, OK, Volume 43, Number 20, September 22, 1993, p. 1.
10. Ibid.
11. LAST, But Not Least: Threats, Costs, & Answers, by Lois Epstein for the Environmental Defense Fund, Washington, DC, February 1993, p. 4.
12. Ibid., p. 5.
13. Shell Building after Massive Chemical Storage Tank Fire, Aboveground Tank Update, Volume 5, Number 7, CEEM Information Services, Fairfax, VA, July 1994, p. 1.
14. 10 Day Fireground Operation at Florida Storage Facility, by Aidan Turnbull, *Industrial Fire Journal,* March and April 1993.
15. Fire Investigation Report of the Denver Tank Farm Fire, by Michael S. Isner, National Fire Protection Association, Quincy, MA, January 1991.
16. Special Data Information Package of Tank Farm Fire Incidents, Fire Analysis and Research Division, National Fire Protection Association, Quincy, MA, 1991.

PART II

Major Regulations of ASTs

CHAPTER 3

The Clean Water Act

The current Environmental Protection Agency (EPA) Oil Pollution Prevention regulations were written in response to acts dating back to the Federal Water Pollution Control Act (FWPCA) of 1972 (PL 92-500). This act dramatically modified and relocated pollution control laws into Chapter 26, Water Pollution Prevention and Control, of United States Code (USC) Title 33, Navigation and Navigable Waters. These laws have been amended a number of times, most notably by the Clean Water Act of 1977 (PL 95-217). It is from the title of this act that the phrase "Clean Water Act" (CWA) is commonly used in reference to the FWPCA and associated amendments to prevent pollution of navigable waters. The CWA laws are contained in 33 USC 1251 et seq.

As stated in 33 USC Section 1251, the purpose of the CWA "is to restore and maintain the chemical, physical, and biological integrity of the Nation's waters." In pursuit of this achievement, the legislation's goals were to:

- Eliminate discharge of pollutants into the navigable waters by 1985.
- Set interim water quality standards for the protection and propagation of fish, shellfish, and wildlife and provide for recreation in and around waters by July 1, 1993.
- Prohibit discharge of toxic pollutants at dangerous levels into Nation's waters.
- Provide federal assistance for the construction of public waste water treatment facilities.
- Implement area-wide waste treatment management planning within each state.
- Initiate a major research effort to eliminate discharge of pollutants into waters.
- Develop policies for the elimination of non-point sources of pollution.

The CWA specifies the role that both the Environmental Protection Agency (EPA) and the states play in implementation and enforcement of the rules, which can overlap. The EPA is required to administer and develop regulations to achieve these goals. However, each state is allowed to develop regulations against pollution of waters within the state that are not superseded by the CWA or associated regulations. The CWA also requires the EPA to develop national standards of performance, enforced by the states, for specific sources, which include chemical manufactures and petroleum refineries, both large users of ASTs. This is important to note because owners and operators of ASTs must examine not only the federal regulations on water pollution but also any applicable state or local regulations. (Chapter 8 discusses various state regulations for the protection of water from pollution discharge.)

Beyond requiring the EPA to implement regulations on the discharge of oil, the CWA mandates penalties and criminal action against any person responsible for an oil discharge. The criminal penalties can be from \$2,500 to \$25,000 or up to one year in prison for a violation resulting from negligence to as high as \$250,000 or up to 15 years in prison for someone who knowingly endangers the public with a discharge of oil. The EPA administrator also can issue civil penalties as a result of a discharge, which can be as high as \$125,000. These figures underscore the importance of understanding the laws and regulations available to prevent the pollution of the nation's waters.

Oil Pollution Act of 1990

Another act designed to prevent the discharge of pollutants into the nation's waters is the Oil Pollution Act (OPA) of 1990 (PL 101-380). This act was a direct result of major oil spills in the late 1980s and was signed into law on August 18, 1990, to provide for worst-case contingency plans for oil spills. Although the CWA protects the waters from the discharge of a number of different pollutants, the OPA was written into law specifically to prevent the discharge of oil into navigable waters from ASTs as well as oceangoing vessels.

The OPA requires the president of the United States to implement regulations against the discharge of oil over land and water from various equipment. Because of the broad range of facilities covered by the OPA, President Bush issued Executive Order 12777 on October 8, 1991, designating the appropriate government agency authority over specific areas. The EPA is required to issue and administer regulations regarding the prevention of oil discharge from non-transportation-related onshore facilities, which are the focus of discussion for this chapter. The Department of the Interior governs offshore facilities, the U.S. Coast Guard covers water transportation-related vessels, and the Department of Transportation Office of Pipeline Safety covers onshore transportation-related facilities.

The OPA was enacted to expand prevention and preparedness activities, improve response capabilities, and ensure that AST owners and operators pay the cost of spills that occur. With this focus, the OPA requires that owners and operators be responsible for a discharge by developing a plan to control a spill, and have adequate liability coverage for potential removal costs. The nation's preparedness was also increased by requiring that the National Contingency Plan (NCP) be revised to include removal of oil and hazardous substances resulting from spills. As part of Executive Order 12777, the president formed a National Response Team (NRT) consisting of representatives from appropriate federal department and agencies. The NRT is responsible for updating the NCP and coordinating the removal of oil and other hazardous material resulting from a spill. The OPA states that the contents of the NCP shall provide for the efficient, coordinated, and effective containment, dispersal, and removal of oil and hazardous discharges and shall include the following partial list of requirements:

- Assignment of duties and responsibilities among the federal agencies in coordination with state and local agencies.
- Identification, procurement, maintenance of equipment and supplies, and proper use of the dispersants.
- Establishment of Coast Guard Strike teams and a national center to carry out the NCP with adequate equipment and personnel.
- Designation of a federal official for each area as assigned by the NRT who shall be the on-scene coordinator.
- Establishment of procedures and standards for removing a worst-case discharge and for the coordination of activities by the various persons responsible for the efforts.

The OPA requires that the president (or the NRT by authority granted it) establish area committees composed of qualified officials of all applicable governments agencies. The purpose of these area committees is to develop Area Contingency Plans (ACP) to coordinate planning for a spill in or near a specific area. The NRT must first approve an area committee and second approve and periodically review the ACP. The ACP should be written with the basic format of the NCP and address any special economic or environmental impact of a spill on an area, as well as coordinate the responsibilities of the governments and the owners of facilities. The NRT also shall conduct unannounced drills in which all parties must participate.

Further discussion of the National response system is included in Chapter 5, in particular the CERCLA section, where specific details of the initial act creating this system (Emergency Planning and Community Right-to-Know Act of 1986) are outlined. The OPA of 1990 simply modified the system to bridge the gaps found in the emergency planning or an oil spill response.

The key concerns of the OPA of 1990 for owners and operators of ASTs are the requirements for Facility Response Plans (FRP) as stated in Section 4202(a)(5). This specifies submittal to the EPA of "a plan for responding, to the maximum extent possible, to a worst case discharge, and to a *substantial threat* of such a discharge, of oil or a hazardous substance." The FRP must:

- Be consistent with the NCP and ACP.
- Identify, and ensure by contract, private personnel and equipment

that can remove a discharge, including a discharge resulting from fire or explosion, and mitigate or prevent the substantial threat of such a discharge.

- Identify the qualified individual having full authority to implement the plan requirements, and require that person's immediate communication with the coordinator, appropriate governmental agencies, and the personnel and equipment required for removal of a spill.
- Describe the training and prevention measures taken at a facility to ensure safety and a response to a discharge, including equipment testing, periodic unannounced drills, and personnel training on response actions.
- Be updated periodically and resubmitted for review after a significant discharge.

The EPA was to issue regulations for FRPs by August 18, 1992. In order for a facility to remain in operation, an FRP was to be submitted to the EPA by February 18, 1993, by the owner or the operator of the AST, who was to operate in compliance with this plan by August 18, 1993. Plans submitted for facilities that, because of their location, could reasonably expect to cause *significant and substantial harm* to the environment by a discharge of oil or a hazardous material had to have the approval of the EPA by August 18, 1993. The EPA may require amendments to the FRP and must periodically review these plans. These facilities can operate for two years after submittal of an FRP without approval, provided that the owner certifies they are operating in accordance with the FRP.

The OPA consolidated and strengthened the elements and limits of liability for the responsible parties of an oil discharge. As stated by Section 1004 of the OPA, the elements of liability include removal costs and damages incurred by the various government agencies or private parties, consistent with the NCP. Damages for oil discharge include those involving natural resources, real or personal property, loss of subsistence, revenue, profits, and earning capacity. The limit of liability for onshore facilities per the OPA is $350 million, which can be decreased to as low as $8 million taking into account the size of a facility, its proximity to navigable waters, and other elements that are relevant to the risk of a discharge. The OPA also allows state and local governments to set more stringent limits and elements of liability. If the owner can prove that the discharge was caused solely by a third party, this party must meet the set liability limits for the discharge. The owners also may use the defense that the discharge resulted from an act of God or war. If the discharge of oil is the result of gross negligence, willful misconduct, or a violation of regulations, there are no limits of liability for the responsible party of a discharge.

In addition to the liability limits of an oil discharge, Subtitle C of the OPA sets stringent civil penalties, which vary according to the amount of spillage and the level of responsibility of a given party. There are a number of different levels of civil penalties, which can be as high as $25,000 per day or $500,000 total for the owner or the operator of a facility. The civil penalties collected are deposited into the Oil Spill Liability Trust Fund, which was authorized by the OPA. This fund is used to pay the difference between the responsible parties' liability limit and the actual removal costs of an oil spill. The existence of these large civil penalties and the liability limits in costs that can be levied against an owner or an operator as a result of an unauthorized oil discharge, demonstrate the importance of implementing a complete and proper FRP to minimize the damage.

As a result of the CWA and the OPA, 33 USC 1251 et seq. has been revised, and the EPA has produced a number of regulations to prevent the discharge of pollutants into navigable waters. The EPA regulations are contained in Chapter 40 of the Code of Federal Regulations (CFR), with the regulations on water quality written in Subchapter D, Water Programs. The regulations in this subchapter prevent discharge of pollutants into waters and set water quality standards. The 40 CFR Parts 110 through 114 are regulations designed to prevent the discharge of oil into navigable waters and will be discussed in detail in this chapter. Of particular importance is Section 40 CFR 112, Oil Pollution Prevention, which contains the regulations to prevent and control a discharge of oil.

Oil Pollution Prevention Regulations

Oil Pollution Prevention regulations mainly consist of requiring AST facilities to prepare a spill prevention, control, and countermeasure (SPCC) plan. Although the regulations are entitled Oil Pollution Prevention, they are commonly referred to as the SPCC regulation because of the SPCC Plan requirement. The SPCC regulation was initially promulgated by the EPA on December 11, 1973, as authorized under Section 311(j)(1)(C) of the CWA or FWPCA. Since the original publication of the SPCC regulations, there have been two minor revisions: on August 29, 1974, and March 26, 1976. On May 20, 1980, the EPA introduced major revisions to the SPCC regulation that would have clarified some provisions in the Oil Pollution Prevention regulations interpreted as discretionary to be mandatory. At that time, discussion by the EPA of the proposed revisions concluded that no modifications were required, and the revisions were not finalized.

After the Ashland tank failure in 1988 (spilling approximately 3.8 million gallons of diesel fuel), the EPA formed the SPCC Task Force to reexamine the SPCC rules.[1] As part of the recommendations, the EPA proposed changes similar to those proposed in 1980. In addition, the task force recommended that technical requirements be added and the scope expanded to include spill contingency plans specific to an oil facility. Because some of the potential changes required further analysis before specific regulations could be proposed, the EPA submitted these revisions in two phases, commonly referred to as Phase I and Phase II. Phase I revisions were proposed on October 22, 1991, and are similar to past proposed revisions. In 1995, the Phase I revisions are expected to be rewritten with changes based on comments received by the EPA. Phase II revisions proposed on February 17, 1993, and finalized on July 1, 1994, included changes required by the Oil Pollution Act of 1990 (OPA).

This section is a combination of the regulations written, including finalized Phase II revisions, and the proposed Phase I changes. The purpose of the current regulations, as stated in 40 CFR 112.1(a) is to "establish procedures, methods, and equipment, and other requirements for equipment to prevent the discharge of oil from non-transportation related onshore and offshore facilities onto or upon the navigable waters of the United States or adjoining shorelines." This phrase offers a world of interpretation by the reader, which makes it essential to define the variables as quantified by the EPA. Facilities, equipment, and operations under the EPA jurisdiction are non-transportation-related onshore and offshore facilities. These facilities are involved in various aspects of oil production, from initial drilling processes to the refining and consumption of oil products. The EPA interprets oil to include crude oil, refined petroleum products, nonpetroleum oil products such as vegetable and animal oils, and oil mixtures containing other substances such as water. Harmful quantities are defined in 40 CFR 110, Discharge of Oil, as oil discharges that violate water quality standards, cause a film or sheen upon or discolor the water surface, or produce sludge or emulsion to be deposited beneath the surface.

Navigable waters was defined in the original regulation and proposed Phase I revisions would bring the definition in line with 40 CFR part 110. The current regulation (Part 110) states that a navigable water way is generally defined as:

- Waters that have been, have the potential to be, or are currently being used for interstate or foreign commerce.
- Interstate waters or wetlands.
- Intrastate water, including lakes, rivers, permanent or intermittent streams, ponds, and a multitude of wetlands used for recreational, commerce fishing, and industrial purposes.
- Tributaries of the above-defined waters.

The definition of navigable waters is broad enough that the majority of facilities in the United States are classified as being near navigable waters. Facilities are under EPA regulations if they could reasonably expect to discharge oil into navigable waters because of their location. The determination is based on the geographical location of navigable waters and excludes artificial deterrents such as dikes or other equipment that may prevent oil from reaching the waters.

AST facilities governed by the EPA SPCC regulations include facilities whose:

- Total aboveground storage tank capacity is greater than 1,320 gallons.
- One-tank capacity is greater than 660 gallons.
- Total underground storage capacity is greater than 42,000 gallons.

The Phase I revisions would include further clarifications on inclusion of underground storage tank (UST) capacity by clarifying the definition of USTs for SPCC planning. USTs as defined in Phase I revisions are tanks that are completely buried and not governed by Section 40 CFR 280, known as the UST regulations. Partially buried tanks, tanks in subterranean vaults, or bunkered tanks are classified as ASTs by the EPA. Partially buried tanks are not recommended but if used should be protected with coatings and cathodic protection because of rapid corrosion, especially at the soil surface. Phase I revisions proposed that the capacity of USTs, at a facility governed by the UST regulations, would not be included in the parameters set to determine whether a facility is governed by these SPCC regulations.

One of the conclusions made by the EPA Task Force was that the EPA does not currently have an adequate inventory of facilities subject to the regulations. Phase I revisions would add 40 CFR 112.1(e), requiring all owners and operators of aboveground storage tanks to submit facility information to the EPA, to include:

- The name of the owner or the operator of the facility.
- The name and the address of the facility.
- The total number and the size of aboveground storage tanks at the facility.
- The total capacity of aboveground storage tanks.
- Distances to the nearest navigable waters.
- Where appropriate and available, the Dun and Bradstreet number and primary Standard Industrial Classification (SIC).

Spill Prevention, Control, and Countermeasure Plans

The major function of the Oil Pollution Prevention (or SPCC) regulations is to require facilities that are expected to discharge oil into navigable waters to maintain a facility SPCC Plan (hereinafter known as the SPCC Plan). The SPCC Plan should be developed by using good engineering practice to prevent a spill or to provide measures to contain and remove a spill following a discharge. Any SPCC Plan developed must be reviewed and certified by a Registered Professional Engineer. The Phase I revisions would expand these requirements specifying that the Professional Engineer shall attest to the following:

- Engineer is familiar with the SPCC regulations.
- Engineer has visited the site and examined the facility.
- SPCC Plan has been prepared according to good engineering practice and SPCC regulations.
- Required testing has been conducted.
- SPCC Plan is adequate for the facility.

The SPCC Plan must be available for on-site review by the regional administrator and be retained at a facility that is normally attended at least eight hours a day, or at the nearest field office. The SPCC Plan not only must be reviewed and certified by a Registered Professional

Engineer but must have the full approval of management at a level possessing the authority to commit the necessary resources for implementation. When there are changes that will affect the facility's potential to discharge oil as a result of construction, maintenance, or operation of the facility, the SPCC Plan must be amended. Additionally, owners or operators are required to review their SPCC Plans once every three years to verify compliance with the SPCC regulations. Within six months of a review, the SPCC Plan must be revised to include prevention and control technology that has proved to significantly reduce the threat of a spill.

If a facility has experienced spills in the past, the regional administrator (RA) may require amendments to the SPCC Plan. A facility that has discharged 1000 gallons or more during a single event or has had two discharges within one year is subject to this provision. Within 60 days of a facility's reporting a discharge, the facility operator shall submit to the RA the SPCC Plan, with complete information on the facility, the cause of the spill, and any corrective action taken to prevent such a spill. The RA, with input from the state, may require that an FRP be revised and adopted within 30 days.

The SPCC Plan requirements discussed in Section 112.7 are often interpreted as discretionary guidelines rather than requirements. Therefore, the SPCC Task Force recommends that the EPA clarify that certain provisions are to be described as requirements rather than guidelines. The original SPCC regulations contained words such as "should" and "guidelines" that some facility owners and operators interpreted as optional. The EPA states that "should" was used to provide flexibility for facilities with unique circumstances that would not apply certain provisions based on good engineering practice. Therefore, Phase I revisions would replace the words "guidelines" and "should" with "requirements" and "shall." The revisions would also divide the requirements of 40 CFR 112.7 into five separate sections to eliminate any further confusion.[2] The general requirements would be in Section 112.7 and the specific requirements for a given facility in sections, 112.8, 112.9, 112.10, and 112.11. Table 3-1 gives the titles of these sections, with a discussion of the requirements to follow in the text. The EPA states that the proposed changes are not intended to be major changes to the SPCC regulation and, by themselves, do not require changes to existing plans. The current regulation will allow variations from the SPCC Plan requirements when they are not feasible or consistent with good engineering practice. The EPA does have the authority to disapprove variations when alternate methods do not provide equivalent protection.

Table 3-1. Proposed Phase I SPCC Plan requirements from 40 CFR 112

Section	Title
40 CFR 112.7	SPCC Plan general requirements
40 CFR 112.8	SPCC Plan requirements for onshore facilities
40 CFR 112.9	SPCC Plan requirements for onshore oil production facilities
40 CFR 112.10	SPCC Plan requirements for onshore oil drilling and workover facilities
40 CFR 112.11	SPCC Plan requirements for offshore oil drilling, production, or workover facilities

Source: EPA October 22, 1991, proposed rule changes to 40 CFR 112.

Section 112.7 would list the general requirements of an SPCC Plan, which can be further divided into the following parts:

- Spill prevention, control, and countermeasure provisions.
- Inspection, tests, and records.
- Brittle fracture.
- Personnel, training, and spill prevention procedures.
- Facility tank car and tank truck loading or unloading rack.
- Security.

According to the Phase I revisions, the spill prevention, control, and countermeasure provisions of the SPCC Plan requirements would be contained in Section 40 CFR 112.7(a)(3) and must include the facility's physical layout. This must include a diagram containing the location and contents of all tanks along with the following:

- Unit-by-unit storage capacity.
- Type and quantity of oil stored.
- Estimates of quantity of oils discharged.
- Possible spill pathways.
- Spill prevention measures, including routine handling of products.
- Spill control such as secondary containment around tanks and other points of a discharge (see Photo 3-1 for an example).
- Spill countermeasures for spill discovery, response, and cleanup.
- Contact list and phone numbers for the facility response coordinator, National Response Center, cleanup contractors,

PHOTO 3-1.
ASTs utilizing secondary containment.

government emergency response teams, and downstream water suppliers that must be contacted in case of a discharge.

Phase I revisions would expand the reporting requirements of a SPCC Plan to include specific information such that a person reporting a spill can adequately explain the spill and its possible consequences. This information should report details of the spill to include location, time, total quantity, and estimated quantity that has reached the navigable waters. Additionally, the spill report should include any damages or injuries; actions that have been taken to stop, remove, and mitigate the effects of the discharge; and whether an evacuation is required. A section of the SPCC Plan should outline the emergency circumstance procedures and be readily available for use, with appendixes included with proper support material.

The SPCC regulation rules require measures of secondary containment and/or drainage control structures or equipment to prevent discharge into the navigable waterways. Phase I revisions would require that the floor and the walls of the entire containment system be impervious to oil for 72 hours, which the EPA regards, in most cases, to be sufficient to detect and remove any discharge of oil. The secondary containment system of an onshore facility shall have one or more of the following prevention measures:

- Dikes, berms, or retaining walls.
- Curbing.
- Drainage systems such as culverts and gutters.
- Barriers such as weirs and booms.
- Spill diversion ponds.
- Retention ponds.
- Sorbent materials.

Section 112.7(d) of the SPCC regulations allow noncompliance with the secondary containment requirements provided that the owner demonstrate such impracticality and written commitment to remove any discharge. Under the Phase II revisions, these requirements would have been strengthened requiring these facilities to submit an FRP in accordance with 40 CFR 112.20 in addition to the SPCC Plan. Although these provisions were not adopted, they may become part of the Phase I revision rewrite.

According to Phase I revisions, inspection of the equipment and the containment system would be required by 40 CFR 112.7(e). These inspections would be completed according to certified procedures, and subsequent inspections signed and maintained within the SPCC Plan for at least five years. Proposed Phase II revisions would have added Section 112.7(i), on evaluation of tanks for brittle fracture. This proposed section was a direct result of the Ashland Oil brittle fracture failure in 1988, to ensure tank integrity after a repair or modification.[3] Following a repair, an alteration, or a change of service of a field-erected AST, a risk evaluation for brittle fracture failure would be completed according to API 653 Section 3. Shop-built tanks would not be subjected to these requirements because of relatively low shell stresses that minimize the risk of brittle fracture failure. Again, these sections were not part of the final Phase II revisions and probably will be included in the Phase I rewrite.

Personnel training is an important part of an SPCC Plan according to the EPA. The Emergency Response Notification System (ERNS) stated in a 1989 report that 12.3% of the oil spills at fixed facilities are related to human error.[4] The cause of the operator error can be a failure to close valves, which can lead to a discharge of oil. The EPA believes the most likely scenario of operator error is the overfilling of a tank during the process of transferring product between tanks. Proper training of operators can help to improve operating and response skills. In addition, training sessions can be used to discuss discharge removal plans and equipment improvement. Therefore, Phase II revisions proposed that 40 CFR 112.7(f) be added, and relate the amount of training required to the size of the facility. Facilities that transfer at least 10,000 gallons in a single transfer twice a month or 50,000 gallons in a single operation are required to hold annual training sessions and participate in unannounced drills. These revisions were also not finalized and should be included in the Phase I rewrite.

Because the EPA holds that discharge of oil occurs mainly during the process of transferring product, Phase I revisions would require loading and unloading racks to meet certain requirements. These would include the design of a catch basin to hold at least the largest-size container to use the rack. The rack area also would have sufficient barriers, either warning lights or physical barriers, to prevent a vehicle from leaving before its disconnection from the facility piping.

Phase I revisions also recommend the addition of 40 CFR 112.7(g), which states the security measures required for a facility to provide protection against the discharge of oil by an accident or by willful misconduct of employees or vandalism. These proposed measures include completely surrounding facilities with fencing and locked gates or guards during times of nonoperation. In addition, valves, start controls on pumps, and loading and unloading connections should be securely locked. A facility must be adequately lighted to enable discovery of a discharge and so that unlawful entry into the facility can be detected at night.

The Phase I revisions would divide the current regulatory requirements of 40 CFR 112.7 into five parts, as shown in Table 3-1. The two most important parts for ASTs would be Sections 112.8 and 112.9, which are the requirements for production and nonproduction onshore facilities. In both cases, the proposed requirements are broken down into three sections: facility drainage, bulk storage containers, and facility transfer operations. The requirements for nonproduction onshore facilities, covered in 40 CFR 112.8, would specify that drainage of rainwater from a diked area shall be restrained by valves or other positive means to prevent accidental drainage of an oil spill. Diked areas containing aboveground storage tanks are required to have a secondary containment sized for the largest tanks plus some freeboard to account for rainwater. Any rainwater should be manually drained. If the rainwater is bypassing the facility treatment system, the water being drained should not contain harmful quantities of oil, as defined by 40 CFR 110, and such events must be recorded. If the area is not diked, adequate techniques of facility drainage should be engineered to prevent an oil spill from entering navigable waters.

According to the proposed Phase I revisions, bulk storage containers or ASTs located in diked areas must be constructed according to a national standard (such as API, ASME, UL) and made of material compatible with and under conditions required by the contents. The ASTs would be subjected to integrity tests at least every ten years and shall consist of any combination of hydrostatic testing, visual inspection, radiograph examination, ultrasonic examination, or other nondestructive testing. New and old tanks would be updated to include one or more of the following measures to prevent discharge:

- High-level alarms with an audible or visual signal at a manned operating station.
- Automatic shutoff valves to stop flow into a tank.
- Direct audible or code signal between the tank level gauge and the pump station.
- A fast response system for determining the liquid level in the tank.
- Regular testing of liquid-level sensors.

For onshore production facilities, the Proposed Phase I SPCC requirements given in Section 112.9 are similar to those discussed for nonproduction facilities except for the following: valves in secondary containment areas shall remain closed unless the water is being drained according to the measures discussed for nonproduction facilities. The regulations would also require that the path of drainage or ditches from a diked area should be inspected regularly to remove any oil within 72 hours. Tanks for production facilities should be visually inspected annually, and the following measures are necessary to prevent overflow of a storage tank:

- Adequate tank capacity to ensure against overfill if tank gauging is delayed.
- Overflow equalization lines between tanks to allow overflow between tanks.

- Adequate vacuum protection to prevent tank failure during a pipeline run.
- Communication between the high-level sensors and the production facility computer.

Although the EPA does not give an exact format of the SPCC Plan, it does cite specific information that must be included. A number of organizations, such as API, produce publications on a recommended SPCC outline and assist the owner or the operator in completing an SPCC Plan. It is important to remember that in addition to EPA requirements for the prevention of pollutant discharge from reaching the waters, state and local governments also have regulations that must be followed.

Facility Response Plans

As required by the OPA of 1990, the EPA implemented Phase II revisions (effective August 30, 1994) to the SPCC regulations, defining the requirements for AST Facility Response Plans (FRP). An owner or an operator of an AST that could reasonably expect to cause *substantial harm* to the environment with a discharge of oil was required to file an FRP with the EPA regional administrator by February 18, 1993, and to operate in accordance with the FRP by August 18, 1993. This FRP could be required in addition to a facility SPCC Plan that had already been implemented. Even though the EPA proposed rules that were not published until February 17, 1993, the February 18, 1993, deadline for submitting FRPs had to be maintained because this deadline was required by the OPA of 1990 and not associated regulations. Those who submitted plans prior to the February 18, 1993, deadline had to resubmit the FRPs for reapproval by February 18, 1995. Facilities that failed to submit an FRP by the February 18, 1993, were required to do so by August 30, 1994. For new facilities or when operational changes occur at a facility, the FRP must be submitted before operations are begun or within 60 days if changes were required during the facility startup. When an unplanned change to a facility occurs that may affect the discharge of oil, the facility has six months to submit the FRP.

The Phase II revisions amended SPCC regulations by adding 40 CFR 112.20, listing the requirements of FRPs. To determine if a facility is required to submit an FRP, the EPA developed a flowchart to classify each facility, which is reproduced in Figure 3-1. This

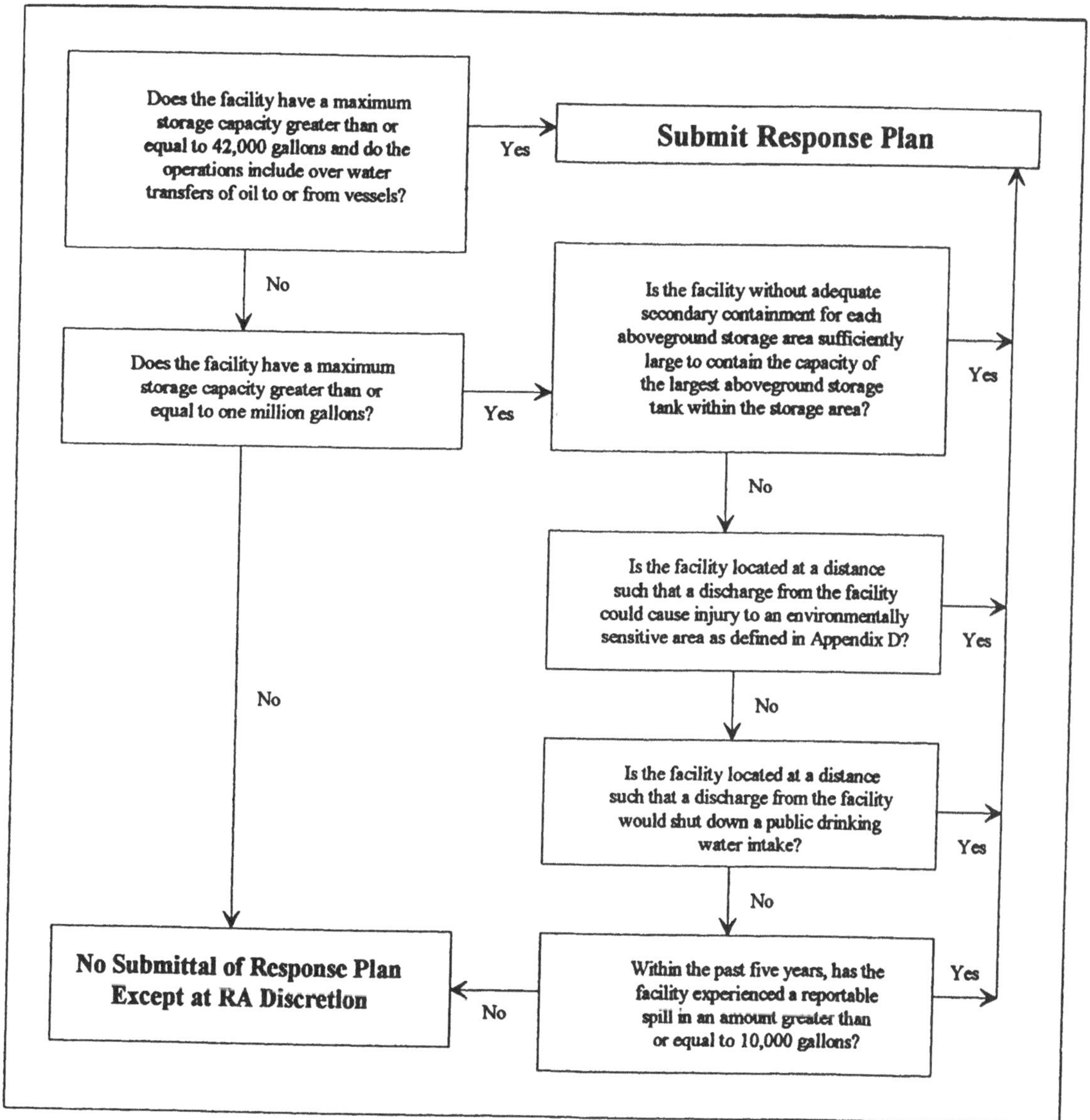

FIGURE 3-1.
Flowchart for the determination of substantial harm (Source: Federal Register)

flowchart represents a self-selection process to determine whether a facility is required to submit an FRP to the RA. The guidelines are based on the size and the location of a facility with respect to navigable waters, and can be separated into two general types of facilities that are required to submit an FRP:

1. ASTs that have a capacity of 42,000 gallons or more and conduct operations over waterways.
2. Facilities that do not operate over waterways but have a total capacity greater than one million gallons and meet certain other criteria.

The latter include facilities that do not have adequate secondary containment, those located at a distance where they could harm fish, wildlife, and environmentally sensitive areas or public water intakes, and facilities that have reported a spill of greater than 10,000 gallons in the past five years. The EPA has defined environmentally sensitive to mean an area identified by either legal designations or by evaluation of the ACP, and this may include areas such as wetlands, critical habitats, sanctuaries, reserves, and national and state parks.

The planning distance between a facility and environmentally sensitive areas or water intakes that would require facilities with a capacity greater than one million gallons to file an FRP is defined in the regulations by Attachment C-III. This regulatory attachment includes a method for determining the planning distance, which is dependent on the oil rate-of-travel over land, streams, or still water and the reaction time to an oil discharge. The time required for an oil discharge to reach a sensitive area should be less than the time required to react to a discharge. The regulations (Table 3 of Attachment C-III) list the reaction time based on recommendations from the U.S. Coast Guard (USCG) and the type of waterway involved. If the waterway is a high-volume port area or the Great Lakes region, the reaction time is 15 hours, whereas in other areas the reaction time is 27 hours. Therefore, if a sensitive area is within the planning distance of a facility, the facility must complete an FRP, given the other parameters discussed in Figure 3-1.

If an aboveground storage tank facility does not pose a threat of substantial harm, the facility is required to file a certification stating that an FRP is not required. However, the RA may require that these facilities submit an FRP according to Section 112.20(b)(1), based on the following criteria:

- Type of transfer operations.
- Oil storage capacity.
- Lack of secondary containment.
- Proximity to environmentally sensitive areas and public water intakes.

- Spill history.
- Other site-specific characteristics and environmental factors.

Upon notice by the EPA of the requirement to file an FRP, the owner has six months to submit a plan.

The OPA of 1990 required the EPA to review each FRP to determine if the facility could reasonably expect to cause a *significant and substantial harm* to the environment by an oil discharge. The FRP was to be approved by August 18, 1993, and subsequently reviewed periodically (which the regulations defined as every five years). The facility could operate without approval after this date for up to two years if the FRP were followed and ensured by contract the required personnel and equipment. If a facility is determined by the RA to cause significant and substantial harm or to make changes that materially affect the implementation of the FRP, the changes must be reported to the EPA within 60 days. Section 112.20(f)(3) lists the guidelines that the EPA may use to determine whether a facility could cause significant and substantial harm to the environment. The RA should consider:

- Frequency of past spills.
- Proximity to navigable waters.
- Age of storage tank.
- Other facility-specific or region-specific information, including health impacts.

The specific requirements of the FRP are given in the SPCC regulations (Sections 112.20(h)). An FRP must be completed consistent with the requirements of the NCP and the ACP. The AST facility owner or operator must review these plans annually and make any necessary amendments. The FRP shall follow the format of the model facility-specific FRP given in the regulations (Appendix F). If an FRP has been prepared to comply with state or other local government regulations, it shall be equivalent to and cross-referenced with the following elements given in 40 CFR 112.2(h):

1. Emergency response action plan.
2. Facility information.
3. Information about emergency response.
4. Hazard evaluation.
5. Tiered planning scenarios.

6. Discharge detection systems.
7. Plan implementation.
8. Self-inspection, drills/exercises and response.
9. Diagrams.
10. Security systems.
11. Response cover sheet.

These elements are also the table of contents to the model facility-specific FRP given in the regulations (Appendix F). The requirements of the FRP are generally very similar to but more stringent than those of the SPCC regulations previously given. Therefore, the discussion of the FRP is brief, and the regulations should be read to gather the precise requirements. Item (3), information about emergency response, is very detailed and contains ten points that must include details of an oil discharge response. The procedures followed to identify response resources to meet the FRP are given in the regulations (Appendix F).

The main focus of the OPA of 1990 and of subsequent Phase II revisions of the SPCC regulations is planning for a worst-case discharge. In regard to item (5), on tiered planning scenarios, the regulations require the FRP to include a discussion of three discharge scenarios. The three cases are:

- *Worst-case discharge,* using the calculations from Appendix D.
- *Small-case discharge,* equal to or less than 2,100 gallons or the worst-case discharge.
- *Medium-case discharge,* greater than 2,100 gallons but less than 36,000 gallons, 10% of the capacity of the largest aboveground storage tank, or the worst-case discharge.

The worst-case discharge determined by the regulations (Appendix D) is based on whether a facility is a single-tank facility or a multiple-tank facility. In general for single-tank facilities, the worst case is the tank capacity or 80% of the tank capacity if proper secondary containment is provided. If all the tanks of multiple-tank facilities do not have secondary containment, the worst-case discharge is equal to the total facility tank capacity. If some tanks do not have secondary containment, the worst-case discharge is equal to the total tank capacity without secondary containment plus the greater capacity of either the largest tank with secondary containment or the total capacity of tanks that are permanently manifolded together. For production facilities, the worst case discharge shall include a capacity based on the production volume of the well.

National Pollutant Discharge Elimination System

In pursuit of the purpose of the CWA, one national goal is to eliminate the discharge of pollutants in navigable waters. To achieve this goal, the CWA (FWPCA of 1972) and the associated regulations implemented in 1973 prohibited point-source pollutant discharge into navigable waters without a proper permit. These regulations have controlled the majority of point pollution from industrial processes and municipal sewage waste, which constitutes the most detrimental and easily identifiable pollution problem. However, these regulations did not adequately control the discharge of pollutants contained in storm water from these sources until the CWA was amended by the Water Quality Act (WQA) of 1987. To eliminate discharge of polluted storm water, a regulation was implemented on November 16, 1990, and is contained in 40 CFR 122, EPA Administered Permit Programs: The National Pollutant Discharge Elimination System or NPDES. The regulations contained in this section are commonly referred to as the NPDES regulations, and the associated permit required is commonly known as the NPDES permit.

The Water Quality Act of 1987 states which type of storm water discharges from a point source requires a NPDES permit, the involvement of the states in the permit process, and the requirements of the storm water discharge permit. The WQA provided a comprehensive framework for the EPA to regulate the discharges of five types of storm water by requiring certain facilities to obtain permits from the EPA by October 1, 1992; these facilities are defined as:

- Those that required a permit prior to February 4, 1987.
- Those associated with industrial activities.
- Those associated with a municipal separate storm sewer system serving a population of 250,000 or more.
- Those associated with a municipal separate storm sewer system serving a population of between 100,000 and 250,000.

The WQA allows the states to apply for administration of the NPDES program instead of the EPA. The EPA must develop procedures for states to follow in developing a program to include guidelines for monitoring, reporting, enforcing, and funding as well as the development of standard permit forms. To date about 40 states have approval to run the permit program. This act also establishes two pollution control standards for conditions in NPDES permits, technology-based and water-quality-based, which are discussed in 40 CFR 125. The two pollutant control technology-based standards appropriate for existing storm water discharges are best conventional technology (BCT) and best available technology (BAT). The water quality standards have more stringent conditions placed on the quality of water being discharged than those of the technology-based standards.

It is important to note which industrial facilities are required to submit a NPDES permit and to understand the role these regulations play with respect to ASTs. The EPA published regulations that require facilities to obtain an NPDES permit for storm water discharges associated with 11 industrial activity categories, grouped together by the Standard Industrial Classification (SIC) code. The first ten are considered heavy industry, and some of the industries that utilize ASTs include petroleum, chemical, paper, mining, and hazardous waste storage. This regulation does not directly affect the installation or the operation of an AST, as was the case for the SPCC regulation; rather the AST market is affected because a facility NPDES permit may require the installation of an AST to control or treat any point-source discharge.

In addition to specifying the industries required to obtain an NPDES permit, the NPDES regulation provides for three different types of permits; individual, group, and general. An individual permit is filed by a single facility for a specific industrial storm water discharge. The permit application should contain information on the facility and spill records of the last three years, as well as the storm water discharge characteristics (i.e. amount of discharge, expected amount of certain pollutants, and oxygen demand). AST owners and operators should note that an NPDES permit is not required for an oil facility if there was not a reportable-quantity spillage in violation of any pollution regulations since November 16, 1987, and the discharge does not violate the water quality standards. A group permit is similar to an individual permit except that an application is submitted in two parts by a group of facilities within the same industrial NPDES classification. The first part of the application contains information on the group members, and the second part lists quantitative data, similar to those required by the individual permit, for a percentage of the group members. Currently, over 1,200 groups with over 60,000 facility members are classified by this type of permit. The third type of permit is the general NPDES permit, which requires a facility to apply by filing a notice of intent (NOI) and to be approved to discharge under a general permit. General permits are produced by the EPA and adhered to by the states and regions that do not administer their own NPDES permit program. The states that do administer an NPDES permit program may modify the general permit for industries within the state.

The EPA estimates that over 100,000 industrial facilities discharge storm water requiring an NPDES permit. This has placed an administrative burden on the EPA and the states to issue and administer the program. Therefore, the EPA has set up a four-tier permit priority system to alleviate the burden. The priorities are:

- *Tier I—Baseline Permitting:* one or more general permits developed to cover the majority of the discharges from storm water associated with industrial activity.
- *Tier II—Watershed Permitting:* individual or watershed-specific general permits for a watershed district that has been shown to be adversely impacted by a facility discharge.
- *Tier III—Industry-Specific Permitting:* individual or industry-specific general permits for facilities in a specific industry classification.
- *Tier IV—Facility-Specific Permitting:* individual permits for specific facilities based on a wide variety of factors.

As part of any permit, monitoring the discharge of storm water is required to ensure that a discharge does not exceed the limits of the NPDES permit. If the discharge continues to exceed the limits of the permit, the storm water may require treatment prior to entering the waters. This may require the installation of a waste treatment facility, which typically necessitates ASTs.

Notes

1. Environmental Protection Agency, Oil Pollution Prevention; Non-transportation-related Onshore and Offshore Facilities; Proposed Rule, 56 FR 54612, October 22, 1991, proposed rule, p. 54613.

2. Environmental Protection Agency, Oil Pollution Prevention; Non-transportation-related Onshore and Offshore Facilities; Proposed Rule, 56 FR 54612, October 22, 1991, proposed rule, p. 54620.
3. Environmental Protection Agency, Oil Pollution Prevention; Non-transportation-related Onshore Facilities; Proposed Rule, 58 FR 8824, February 17, 1993, proposed rule, p. 8842.
4. Environmental Protection Agency, Oil Pollution Prevention; Non-transportation-related Onshore Facilities; Proposed Rule, 58 FR 8824, February 17, 1993, proposed rule, p. 8841.

CHAPTER 4

The Clean Air Act

The increase in air pollution brought on by urbanization and industrial development endangered the public health and welfare and prompted Congress to establish the first Clean Air Act (CAA) in 1955. The purpose of the CAA is to protect and enhance air quality, by mandating that state and local governments be primarily responsible for air pollution prevention and by accelerating national research and development programs. Aboveground storage tanks must meet the requirements of the Clean Air Act regulating air pollutant emissions into the atmosphere. The current legislation exists under Title 42, Chapter 85 of the United States Code (USC). The CAA has been amended a number of times, most notably in 1970, 1977, and 1990, with this latest revision promulgating a number of pertinent changes to the act.

As part of the CAA, the Environmental Protection Agency (EPA) is required to publish a list of pollutants that cause and contribute to air pollution. Within 12 months after placement of a pollutant on the list, the EPA must issue air quality criteria outlining possible detrimental effects on public health and welfare caused by its emission into the ambient air. The EPA then must prescribe primary and secondary National Ambient Air Quality Standards (NAAQS), which set levels for maximum concentrations of pollutants in the atmosphere. Primary standards are set to protect public health, and secondary standards are to protect public welfare from any known or anticipated adverse effects associated with the presence of the pollutants.

Another responsibility of the EPA under the CAA is the issuance of Control Techniques Guidelines (CTGs), which are used by states in their formation of State Implemented Plans (SIP) to control emissions from existing sources in areas that do not meet the ambient air quality standards. CTGs should be issued simultaneously with the air quality criteria and contain information on the type of control needed, cost of installation and operation, emission reduction, and environmental impacts. The federal government generally does not get involved in the implementation and enforcement of the emission controls, but is responsible for assisting the states by issuing the CTGs. Prior to November 15, 1990, the EPA had not issued CTGs for 11 categories of stationary sources of volatile organic compounds (VOC). However, the 1990 CAA amendment (42 USC Section 7511B) mandated that CTGs be written for these stationary emission sources of VOC by November 15, 1993. This section also mandated that the existing CTGs be reviewed and updated as required, and that alternative control techniques be available for emissions from all sources emitting 25 tons per year of VOC or nitrogen oxides (NO_x) during the same three-year period.

The governor of each state must designate to the EPA specific areas that either meet or do not meet the primary and secondary NAAQS within their jurisdiction. Within one year of the EPA's issuing an NAAQS for a specific pollutant, the state must designate these areas as nonattainment, attainment, or unclassifiable for that pollutant. Nonattainment means that the area does not meet the requirements of the NAAQS for a certain pollutant. Conversely, attainment means that the NAAQS requirements have been met. Unclassifiable refers to an area that cannot be classified as either attainment or nonattainment based on the available information. The state is responsible for the designation of all areas. However, if the state fails to designate an area or does so incorrectly, then the EPA can and will provide an appropriate classification.

State Implementation Plan

Beyond area designation, the state must adopt a State Implemented Plan (SIP) for maintaining and enforcing the NAAQS for a given pollutant. A change to the SIP must be initiated within three years after the EPA implements a new NAAQS. The requirements for an SIP, as indicated in 42 USC Section 7410, must include emission limitations, controls measures, and timetables for compliance with the NAAQS for a given air pollutant.

For attainment areas, the requirements for the SIP are given in 42 USC Part C, termed the "Prevention of Significant Deterioration" (PSD) of air quality. The major portion of Part C is the regulation of new and modified major sources of pollutants in attainment areas. The EPA defines a major stationary source as one that does emit or has the potential to emit 100 tons per year of any one pollutant from a source listed in the CAA. Some of these sources listed under 42 USC Section 7479 include petroleum refineries and storage and transfer facilities. The CAA also states that any unlisted source is considered a major source if the emissions are 250 tons per year or more for any air pollutant. The owner or the operator of a new major source is required to implement Best Available Control Technology (BACT) for each pollutant and to obtain permits for construction of a new source, such as a tank, that must prove that the source will not alter the state of attainment for a given area. The BACT requirements for a specific source should be stated in the SIP.

For areas that are considered nonattainment, the requirements for the SIP are given in 42 USC Part D. Each SIP must be revised to include several requirements stipulated by the CAA to reach attainment. First, the plan must include provisions for the application of Reasonably Available Control Technology (RACT) standards for existing sources as expeditiously as practical when required. Second, the plan also must demonstrate Reasonable Further Progress (RFP) in the annual incremental reductions of air pollutants to ensure attainment of an area by the required date. Next, each SIP must contain provisions for reporting to the EPA on the inventory of sources in an area and also for permits to control the addition of new sources. The inventory of sources must be an accurate and detailed account of the emissions within an area and was to be submitted to the EPA within two years after November 15, 1990, with periodic updates to follow every three years. As part of the investory report to the EPA, owners and operators of sources that emit VOC or NO_x were required to submit emission statements by November 15, 1993, and every year thereafter. States may waive this requirement for sources that emit less than 25 tons per year. Permit requirements for new sources can be found in 42 USC Section 7503. This section states that the proposed source must comply with the Lowest Achievable Emission Rate (LAER). LAER controls must reflect the most stringent emission limits contained within the SIP or limits that are achievable in practice for a given classification of source, but in no case should the source emit any pollutant in excess of that allowed by the New Source Performance Standards (NSPS). To obtain a permit, the owner or the operator also must demonstrate that all other major sources are subject to and comply with the state emission limits allowed by the SIP. Finally, the SIP must include qualifications for the offset provisions allowed by the CAA. If a new or a modified source under consideration by permit will increase the emissions of an air pollutant controlled by NAAQS, the owner or the operator may implement what the EPA terms "offset requirements." By this method, the increase in emissions from a source must be offset by a reduction in emissions of the same pollutant at the same source or an alternative source in the same nonattainment area. The offset can be taken in other nonattainment areas if given prior state approval. The amount of offset is dependent upon the source classification for a given pollutant and is provided as a ratio of emissions decrease to emissions increase, which must always be greater than one.

The air pollutants of most relevance to aboveground storage tank owners are volatile organic compounds (VOC) and nitrogen oxides (NO_x). When combined with sunlight, VOC and NO_x form ozone (O_3), an air pollutant for which both primary and secondary NAAQS exist. The CAA promulgates requirements for SIPs to reach attainment in areas designated as nonattainment for ozone under the NAAQS. These requirements are given in 42 USC Section 7511, which further classifies nonattainment areas as marginal, moderate, serious, severe, or extreme, according to the severity of the ozone pollution. The classifications, required attainment dates, and offset requirements for RACT implementation on major sources are given in Table 4-1.

Table 4-1. Ozone nonattainment classification

Nonattainment Area Classification	Ozone Level	Attainment Date[1]	Defined Major Source (tons/yr.)	Offset Ratio[2]
Marginal	0.121 up to 0.138	1993	100	1.1 to 1.0
Moderate	0.138 up to 0.160	1996	100	1.1 to 1.0
Serious	0.160 up to 0.180	1999	50	1.2 to 1.0
Severe	0.180 up to 0.280	2005	25	1.3 to 1.0
Extreme	0.280 and above	2010	10	1.5 to 1.0

[1] Attainment date is November 15 of the year given.

[2] Emission offset ratio is equal to the total reduction to the total source increase of ozone.

Another notable requirement of the SIP, beyond the provisions listed in Table 4-1 to ensure future attainment of the ozone NAAQS, is the implementation of VOC emission reduction. There are no VOC emission reduction requisitions for marginal areas. For moderate areas, the SIP was to be revised within three years after November 15, 1990, to provide for a 15% decrease in VOC emissions by the attainment date and within two years to allow for the implementation of RACT standards to all major sources of VOC emissions and each category of VOC sources covered by CTGs until the time of attainment. For the remaining classifications, the SIP was to be revised by November 15, 1994, with provisions added to obtain a 15% reduction in VOC emissions for the period ending November 15, 1996. The plan must also provide at least a 3% reduction every year during this period. The addition of NO_x controls to the plan could be substituted for the amount of VOC emissions required.

For severe and extreme areas, the CAA implemented strict penalties for the failure to comply with the requirements. These penalties, which are listed in 42 USC Section 7511D, outline required fees to be paid to the state by owners and operators, and sanctions brought upon the state for failure to enforce the SIP.

New Source Performance Standards ///////////////

New Source Performance Standards (NSPS) are controls placed on new sources only and reflect the best system of emission reduction of a specific air pollutant as determined by the EPA. As part of the 1970 amendment to the CAA, the EPA must list categories of stationary sources and implement standards of performance within one year of specifying the air pollutant. It is the states' responsibility to implement and enforce these NSPS, as is the case also for RACT standards in nonattained areas. The difference between RACT standards and NSPS are that NSPS are required for new sources in any area, whereas RACT standards involve only existing sources in nonattainment areas.

The NSPSs that the EPA has written to control emission of VOC and NO_x from new sources are given in the Code of Federal Regulations (CFR). Those sections most relevant to owners and operators of ASTs listed in 40 CFR 60 are Subparts K, Ka, and Kb, which are the NSPS for tanks storing petroleum liquids and warrant further discussion.

K Controls

The first part of the EPAs regulations listed in 40 CFR 60 for the storage of petroleum liquids is Subpart K. This section defines petroleum liquids as any finished or intermediate products manufactured in a petroleum refinery excluding Nos. 2 through 6 fuel oils, gas turbine fuels, or diesel fuels. The regulation also defines storage of these fuels as covered by Section 60.17. The first regulation of VOC emissions from tanks storing petroleum liquids was mandated under the 1977 Clean Air Act for tanks built, reconstructed, or modified between June 11, 1973, and May 19, 1978. Tanks smaller than the 40,000-gallon limit or pressure vessels with an operating pressure greater than 15 psig are not governed under these regulations.

The type of vapor loss control on a tank is dependent upon the capacity and the true stored product vapor pressure. The true vapor pressure is defined by the equilibrium partial pressure exerted by a petroleum liquid as determined by methods listed in the American Petroleum Institute (API) Bulletin 2517, Evaporation Loss from External Floating Roof Tanks. This pressure differs from the absolute vapor pressure or the Reid vapor pressure, which is listed in the American Testing Society of Materials (ASTM) D323-82 for crude and nonviscous petroleum liquids. Section 40 CFR 60.112 lists the standards for the control of VOC losses from storage tanks. This regulation requires that tanks storing petroleum liquids with a true vapor pressure equal to or greater than 1.5 psia but not greater than 11.1 psia shall be equipped with vapor control equipment such as a floating roof, a vapor recovery system, or its equivalent. The EPA defines a floating roof as a double deck, pontoon single deck, or internal (covered) floating roof supported by the stored liquid. Subpart K also requires that a floating roof be equipped with a closure seal between the roof and the tank wall. When the true vapor pressure is greater than 11.1 psia, the tank may only be equipped with a vapor recovery system.

The owners or the operators of a petroleum liquid storage tank are required to monitor operations and to maintain records indicating the type of liquid stored, the period of storage, the temperature, and the maximum true vapor pressure. This is required for any tank, with or without a control device, storing a petroleum liquid with a maximum and absolute vapor pressure greater than 1.0 psia.

Ka Controls

The second part of the EPA regulations required by the Clean Air Act of 1980 is contained in Subpart Ka (40 CFR 60.112a). Tanks constructed, reconstructed, or modified between May 18, 1978, and July 23, 1984, must contain controls covered by this subpart. The size of tanks requiring controls per this regulation did not change from Subpart K and is still pertinent for tanks greater than 40,000 gallons.

Subpart Ka expands upon the requirements given in Subpart K, which govern the storage of petroleum liquids with a vapor pressure equal to or greater than 1.5 psia but not greater than 11.1 psia. This section states that these storages tanks are required to have one of the following controls: an external floating roof for an open top tank or an internal floating roof or a vapor recovery system for fixed roof tanks. These controls and the vapor pressure limits are identical to those of Subpart K.

For external floating roofs, the roof must be a pontoon type or a double deck type and is required to float on the liquid at all times except during the initial filling and emptying function, when the floating roof is resting on support legs. The EPA states that this initial filling and emptying function should be continuous and accomplished as rapidly as possible.

The closure device between the floating roof and the shell consists of two seals, one on top of the other. The lower seal is defined as the primary seal and the upper seal as the secondary seal. The primary seal may be either a metallic shoe, liquid-mounted, or vapor-mounted. A metallic shoe seal is defined as a metal sheet held vertically against the tank wall by springs or weighted levers. One end of the shoe must be in the liquid with the other end extending a minimum of 24 inches above the liquid level. Liquid-mounted seals are filled with foam or liquid and are mounted in contact with the liquid surface. A vapor-mounted seal is a foam-filled seal that is mounted above the liquid level. The secondary seal for an external floating roof is to be located above the primary seal and should completely cover the space between the tank shell and the floating roof. Subpart Ka further defines external floating roof seals by limiting the gaps allowed between the tank shell and the floating roof. For primary seals mounted in the product, the accumulated area of gaps between the seal and the shell cannot exceed 10.0 square inches per foot of tank diameter with the maximum gap between the shell and the seal not exceeding 1.5 inches. For primary seals that are vapor-mounted, the cumulative area of gaps between the seal and the shell is further reduced to 1.0 square inch per foot of tank diameter. The allowable gap limits for the secondary seal are dependent upon the type of primary seal installed. For a liquid-mounted primary seal, the gap between the secondary seal and the tank wall is limited to 1.0 square inch per foot of tank diameter, with the maximum gap equal to 1.5 inches. For vapor-mounted primary seals, a gap cannot exist between the shell and the secondary seal. The regulations also state that the primary and secondary seals should not contain any holes, tears, or other openings.

For fixed roof tanks, the first type of control is an internal floating roof. For internal floating roofs, the EPA simply states that there should be a closure device between the shell and the floating roof. The regulations give no requirements for the primary seal and do not require that the floating roof have a secondary seal.

Subpart Ka also states requirements for the apertures on internal and external floating roofs. Each opening in the floating roof must extend below the liquid surface and be equipped with a cover, seal, or lid to be maintained in the closed position at all times. Floating roof vents are excluded from this requirement. Automatic bleeder vents are required to be closed when the floating roof is in operation.

The second type of control, for a fixed-roof tank, is a vapor recovery system. A vapor recovery system is also the only control allowed in the storage of a petroleum liquid with a true vapor pressure greater than 11.1 psia. A vapor recovery system is required to collect and to process all VOC vapors and gases discharged from tanks with an efficiency of at least 95% by weight.

In addition to mandating the design of controls for petroleum tanks, Subpart Ka also details the required operation and maintenance of the floating roof. Once the seals for the floating roof are installed, the owner or the operator is required to measure the gaps between the seals and the tank wall. The measurements should be taken every 60 days after first fill for both the primary and secondary seals, at least every five years thereafter for the primary seal, and at least every year thereafter for the secondary seal. This maintenance clock starts after the first fill when the tank is first put in service or after the tank has been out of service longer than one year. The measurement of gaps should be taken at one or more floating roof levels while the roof is floating. The gaps between the seals and shell should be measured when a 1/8″-diameter rod can be passed freely between the seal and the shell. The cumulative area then should be measured by using probes of various widths, multiplied by the circumferential distance of the gap. If the gaps do not fall within the limits, a report must be furnished to the administrator within 60 days of the test outlining the corrective action planned to bring the gaps between the shell and the seal within acceptable limits.

Kb Controls

The EPA's third control requirement for VOCs initiated by the 1987 amendment of the Clean Air Act is the addition of Subpart Kb. This subpart governs any tank constructed or modified after July 23, 1984, and governs tanks currently being constructed.

Subpart Kb lowers the vapor pressure requirements on tanks and the product stored. Tanks with a capacity of greater than or equal to 10,500 gallons (40 m^3) storing a volatile organic liquid (VOL) and having a true vapor pressure of 2.2 psia (15.0 kPa) or greater and tanks with a capacity greater than or equal to 40,000 gallons (151 m^3) storing VOL and having a true vapor pressure of 0.5 psia (3.5 kPa) or greater are required to meet the provisions of this subpart. These limits on capacity and true vapor pressure are different from those in the previous subparts on VOC control because of the recordkeeping requirements of this subpart, discussed later. The provisions of subpart Kb do not apply to a number of tank categories including pressure vessels designed to operate in excess of 30 psig (204.9 kPa), vessels located at bulk gasoline plants, and tanks at gasoline service stations.

Vapor loss controls are not required on all tanks regulated under this subpart as described above. In accordance with Subpart Kb, control of vapor loss is required on tanks with a capacity of 40,000 gallons (151 m^3) or greater storing VOL with a vapor pressure of 0.75 psia (5.2 kPa) or greater but less than 11.1 psia (76.6 kPa). For tanks with a capacity greater than or equal to 20,000 gallons (75 m^3), the lower limit on the true vapor pressure of the product is increased to 4.0 psia (27.6 kPa). Tanks that fall into this category shall be equipped with VOC emission control devices allowed by the EPA. The type of controls may be an internal floating roof, a vapor recovery system, an external floating roof, or an alternate system approved by the EPA. Table 4-2 summarizes the capacity and pressure levels that require different types of controls.

Table 4-2. VOC vapor Kb control limits

Tank capacity, gallons (m^3)	Product vapor pressure (vp), psia (kPa)	Type of control allowed[1]
20,000 (75) $\leq$ capacity < 40,000 (151)	4.0 (27.6) $\leq$ vp < 11.1 (76.6)	IFR, EFR, VRS
	vp $\geq$ 11.1 (76.6)	VRS
capacity $\geq$ 40,000	0.75 (5.2) $\leq$ vp < 11.1 (76.6)	IFR, EFR, VRS
	vp $\geq$ 11.1 (76.6)	VRS

Source: 40 CFR 60 Subpart Kb
[1]IFR: internal floating roof; EFR: external floating roof; VRS: vapor recovery system.

Tanks with a capacity of greater than 10,500 gallons (40 m^3) must maintain records stating the capacity, product stored, and true vapor pressure. If, based on the capacity and the product originally stored, a tank does not fall within the control limits set forward in this subpart, but the true vapor pressure of the product changes such that controls are required, the administrator of the EPA must be notified within 30 days of the change in tank status.

There are three types of controls allowed by the EPA to reduce emissions in Subpart Kb (commonly referred to as Kb controls): an internal floating roof, a vapor recovery system, or an external floating roof. The EPA requires that a fixed roof tank be equipped with either of the first two types of controls: an internal floating roof or a vapor recovery system. An internal floating roof should rest on the liquid surface but necessarily be in complete contact with the surface. The floating roof can rest on the support legs during the filling or emptying of the tank, which should occur as continuously and rapidly as possible. An internal floating roof should be equipped with a closure seal between the tank shell and the floating roof. Three types of seals are allowed by the EPA: a liquid-mounted seal filled with foam or liquid, two seals mounted one on top of the other with the lower seal either a liquid- or a vapor-mounted seal, and a mechanical shoe seal. A mechanical shoe seal may be held against the shell with either springs or weighted levers connected to the floating roof. A coated fabric should span the area between the metal plate and the floating roof.

The second type of VOC vapor control system allowed for a fixed roof tank is a closed vapor recovery system. A vapor recovery system is required to collect and process all VOC vapors and gases discharged from tanks with an efficiency of at least 95% by weight. A vapor recovery system is the only control allowed in the storage of a petroleum liquid with a true vapor pressure greater than 11.1 psia and a tank capacity of 20,000 gallons (75 m^3).

The third VOC control type, an external floating roof (a pontoon or a double deck cover), also should rest on the liquid. While the floating roof rests on legs, the process of filling and emptying the tank should be done as continuously and rapidly as possible. External floating roofs are required to have two seals one on top of the other, or primary (lower) and secondary (upper) seals. The primary seal must be either a mechanical shoe or a liquid-mounted seal. The requirements for the mechanical shoe seal are the same as for an internal floating roof except the plate against the shell must extend into the product and 24.0 inches above the product level. The secondary seal is required to cover the gap between the floating roof and the shell (annular space). This subpart eliminated vapor-mounted primary seals, which were allowed in Subparts K and Ka.

In addition to seals, the EPA mandates control of apertures on the floating roof. Autobleeder and rim vents for floating roofs shall be gasketed and operate only while the floating roof is resting on legs. Rim vents can operate while the roof is floating, venting the annular space at the manufacturer's recommended settings. All other openings in floating roofs must extend into the product, and, except for drains, leg sleeves, and tank penetrations of the deck, shall be gasketed, covered, and maintained in the closed position when not in use. Tank penetrations of the deck are column and ladder wells, which the EPA mandates be gasketed to reduce the gap between the penetration and the floating roof. Sampling wells for internal floating roof tanks and emergency drains on external roofs must be covered with a slotted membrane fabric that covers 90% of the opening area.

As in prior subparts, the EPA mandates periodic inspection of the VOC control system. For internal floating roofs, the floating roof and seals shall be inspected for defects or openings. An inspection should be made after installation, with any necessary repairs or modifications to the floating roof occurring prior to the first fill of the tank. Subsequent inspections of the floating roof should be made from the fixed roof manholes or vents every 12 months after the first fill. If the floating roof is not resting on the product, if defects in the seal are discovered, or if product exists on top of the floating roof, the tank must be removed from service within 45 days, and a report must be filed with the EPA within 30 days. A 30-day extension on removal of the tank from service may be granted if just cause is demonstrated.

A complete inspection of an internal floating roof shall be performed every time the tank is emptied and degassed. The seals and the floating roof should be closely inspected to ensure compliance with this subpart, and any repairs should be made prior to the refilling of the tank. The frequency of these inspections is dependent upon the type of annular space seals utilized. If the tank has both primary and secondary seals, inspection of the floating roof and seals should be done within a five-year interval. For tanks with only a primary seal, this inspection should be done within a ten-year interval. The EPA must be notified at least 30 days in advance of refilling to afford it the opportunity to have a representative present for the inspection. A report of the inspection must be filed with the EPA by the owner or the operator within 30 days after its completion.

The inspection of an external floating roof is twofold. First there is a complete visual inspection of the seals every time the tank is emptied and degassed. The floating roof and seals should be visually inspected to ensure that tears, holes, or other openings do not exist in the floating roof. The EPA must be notified 30 days prior to refilling for a possible inspection.

The second part of the inspection involves the measurement of gaps between the seals and the tank shell. The measurements should be taken for both the primary and the secondary seals during the hydrotest or 60 days after first fill. Subsequent gap measurements of the secondary seal are required annually. The primary seal gap measurements should be taken when the tank is emptied and degassed, at intervals of no more than five years. This maintenance clock restarts after the tank has been out of service longer than one year.

The measurement of gaps should be taken at one or more floating roof levels while the roof is floating. The gaps between the seals and the shell should be measured when a 1/8″-diameter rod can be passed freely between the seal and the shell. The cumulative area then should be measured by using probes of various widths, multiplied by the circumferential distance of the gap. The acceptable gap measurements for external floating roof seals are nearly identical to those listed in Subpart Ka. For primary seals, the accumulated area of gaps between the seal and shell cannot exceed 10.0 square inches per foot of tank diameter, with the maximum gap between the shell and the seal not exceeding 1.5 inches. The gap between the secondary seal and the tank wall is limited to 1.0 square inch per foot of tank diameter, with the maximum gap equal to 0.5 inch. The EPA must be notified 30 days in advance of measurement of the gaps to give the administrator an opportunity to send a representative to the inspection. A report of the gap measurements, including the raw data and calculations, should be submitted to the EPA within 60 days. If the gaps do not fall within the required limits, this report shall be furnished within 30 days of the test. The report also shall describe the corrective action planned to bring the gaps between the shell and the seal within acceptable limits. The seals must be repaired within 45 days of the inspection, with a possible 30-day conditional extension.

National Emission Standards for Hazardous Air Pollutants

Prior to 1990 the Clean Air Act initiated controls on Hazardous Air Pollutants (HAPs) that were found to be detrimental to the public health and welfare. The EPA issued a number of emission controls based on this scheme under 40 CFR 61, with the most applicable being benzene emissions from storage tanks as covered by Subpart Y. The 1990 amendment to the Clean Air Act (CAA), specifically Title III, changed the format of promulgating the regulations of HAPs. The CAA (42 USC Section 7412) currently lists 189 HAPs and mandates the EPA to implement controls on both new and existing sources in a given category to reduce their emissions. In addition, the EPA must examine further controls in order to protect the public health and welfare.

The EPA published a list of all categories and subcategories of major and area sources on July 19, 1992 (57 FR 31576), which had been required under the CAA by no later than November 15, 1991, and must be revised every eight years as appropriate. A major source is considered any stationary or group of stationary sources that emits or has the potential to emit 10 tons per year of any HAP or 25 tons per year of any combination of HAPs. An area source encompasses any source not considered a major source. By November 15, 1995, the EPA must list sufficient area sources, as a source category, to ensure that 90% of the area source emissions of the 30 most harmful HAPs are covered by EPA controls.

In addition to defining categories, the CAA also mandates the time frame in which the EPA must implement controls on new and existing sources. The EPA was required to adopt and circulate emission controls on the first 40 categories within two years of November 15, 1990, on 25% of the listed categories in four years, and on 25% more within seven years of November 15, 1990. Emission standards for all categories must be adopted by November 15, 2000. As required by the CAA, the EPA published a schedule for promulgating source category standards (December 3, 1993, 58 FR 63941) based on these dates. Should the EPA fail to adopt standards for a given source by the date previously published, the owner or the operator of the source must submit an application pursuant of the state permit program within 18 months of the scheduled standard promulgation. As part of the CAA, states were to have permit programs in place by May 15, 1994, containing emission limits of HAPs that would enable the states to administer the self-imposed emission limits of HAPs, if required. A state program must be approved by the EPA before an owner or an operator of a source could implement any emission limitations on HAPs.

The emission standards prompted by 42 USC Section 7412 must limit the emissions of the HAPs to the Maximum Achievable Control Technology (MACT). The MACT standards differ between new and existing sources and are based upon the number of sources under a category. For new sources, the degree of control for a source should be the emission controls achieved in practice by the best controls in similar sources, as determined by the administrator. MACT controls for existing sources should not be less stringent than the average of the best performing 12% of existing sources. For categories with fewer than 30 sources, the MACT controls for existing sources should be at least as stringent as the best-performing 5 sources.

Hazardous Organic NESHAP

The EPA has issued regulations (59 FR 19402, April 22, 1994) that will control emissions of 149 Hazardous Organic NESHAPs (commonly referred to as HON) from the category Synthetic Organic Chemical Manufacturing Industry (SOCMI). Prior to the HON rule, the EPA had issued a number of regulations that affected the SOCMI, in particular VOC and benzene emissions from storage tanks. This regulation will provide for a more comprehensive control of emissions within the SOCMI and also cover equipment leaks of seven other processes from sources including styrene/butadiene rubber production, chlorine production, pesticide production, chlorinated hydrocarbon use, pharmaceutical production, and miscellaneous butadiene use. To define the SOCMI source category, the EPA listed 396 synthetic organic chemicals that can result in emissions of HON from sources. The regulation defines a source as a process vent, storage tank, transfer tank, wastewater stream, or equipment leak. Only processes that use or produce one or more of the HAPs listed are subject to these regulations.

The HON regulation sets controls or Referenced Control Technology (RCT) on sources to limit the emissions of HAPs. The RCT applies only to Group 1 sources, which the EPA defines as part of the regulation. The application of RCT controls on a source is dependent upon the amount of emission reduction required at each source or group of sources. The reduction in emissions required for a source is equivalent to the RCT applied to all Group 1 emission points in the group of sources. The owner or the operator of a source can use two methods for achieving this reduction: apply the RCT to each of the sources or average the emissions from two or more sources.

For storage tanks, Group 1 sources are defined by the true vapor pressure, the capacity of the tank, and whether the source is existing or new. Group 1 existing sources are defined as tanks with a capacity of 40,000 gallons (151 m^3) or greater storing a liquid with a vapor pressure of 0.75 psia (5.2 kPa) or greater but less than 11.1 psia (76.6 kPa). For tanks with a capacity less than 40,000 gallons (151 m^3) but greater than or equal to 20,000 gallons (75 m^3), the lower limit on the true vapor pressure of the product is increased to 1.9 psia (13.1 kPa). For Group 1 new sources, the vapor pressure and tank capacity limits are more stringent than for existing sources and are defined for tanks with a capacity of 40,000 gallons (151 m^3) or greater storing a liquid with a vapor pressure of 0.1 psia (0.7 kPa) or greater but less than 11.1 psia (76.6 kPa). For tanks with a capacity less than 40,000 gallons (151 m^3) but greater than or equal to 10,000 gallons (38 m^3), the lower limit on the true vapor pressure of the product is increased to 1.9 psia (13.1 kPa).

All Group 1 storage tanks require RCT emission control devices that are modeled after the Kb controls of the NSPS. The type of controls allowed are either an internal floating roof, an external floating roof, an external floating roof converted to an internal floating roof with the addition of a fixed roof, or a closed vent system with a control device. These proposed RCTs for emission control are described briefly below, with only key points and major variances from the Kb controls discussed.

An internal floating roof is required to have a closure device between the floating roof and the tank shell that may be one of the following: a liquid mounted seal filled with foam or liquid; two seals, primary and secondary, with the primary being either a liquid- or vapor-mounted seal; or a mechanical shoe seal.

An external floating roof must be a pontoon or a double deck cover and is required to have both a primary and a secondary seal. The primary seal must be either a mechanical shoe or a liquid-mounted seal. The secondary seal is required to cover the gap between the floating roof and the shell (annular spce). As the HON rule applies to both existing and new vessels, existing seals will be allowed to remain for a period of time if they do not presently conform to these requirements. Specifically, if the existing seal installed prior to December 31, 1992, is a single primary liquid-mounted seal or a vapor-mounted primary seal with a secondary seal, the floating roof seals must be brought into conformance the next time the roof is emptied and degassed, or by April 22, 2004. The aperture requirements for an external floating roof are identical to the VOC emissions Kb controls for an external floating roof with the exception of the gauge pipe. The gauge pipe is required to have a gasketed sliding cover between the floating roof and the pipe and also must be sealed on the inside of the pipe to reduce emissions. These requirements vary, depending upon whether or not the pipe is slotted. If the gauge pipe is slotted, the pipe must have a gasketed float inside the pipe.

When an external floating roof is converted to an internal floating roof, the floating roof must comply with the requirements of an internal floating roof, and the floating roof fittings must adhere to the requirements associated with external floating roofs. This would mean that only a liquid primary seal would be required for the closure between the tank shell and the floating roof, and that the gauge pipe should be equipped with a float if slotted.

The last type of HAP vapor control system allowed is a closed vent system with a control device for a fixed roof tank. A closed vent system is required to collect all HAP vapors and gases discharged from tanks with a maximum emission rate of 500 parts per million above background detection. The control device should be designed to reduce emissions by at least 95% by weight. A vapor recovery system is the only control allowed in the storage of a petroleum liquid with a true vapor pressure greater than 11.1 psia (76.6 KPa) and a tank capacity of 20,000 gallons (75 m^3).

Gasoline Distribution NESHAP

The EPA has proposed NESHAPs (issued on February 8, 1994) for gasoline distribution, which was listed in the EPA's initial list of source categories. For purposes of the proposed regulation, the gasoline distribution network refers to the storage and the transfer of gasoline as it moves from the refinery to service station pumps. An analysis of this system reveals that there are five distrinct subcategories: pipeline pumping stations, pipeline breakout stations, bulk gasoline terminals, bulk plants, and service stations. To begin, gasoline is carried from the production facility to terminals by pipelines, which can be miles in length. When the pipeline length increases to a point at which the initial pump cannot move the product sufficiently, pumping stations are installed to aid the product flow. Occasionally, the flow may be diverted into storage tanks or breakout points, which usually coincide with pumping stations. Bulk gasoline terminals receive the product from the pipeline, or by other means of transportation, and store it for distribution. From this point, the product is delivered either to service stations or to intermediate storage and handling facilities known as bulk plants.

Under the CAA, all new and existing major sources in a given source category must conform to MACT standards. Of the gasoline distribution source category, pipeline breakout stations and bulk terminals are the only two that have been found to contain major source facilities. Therefore, the proposed regulations only apply to these two subcategories.

The owner or the operator of a pipeline breakout station or bulk terminal must determine if the facility is required to conform to the regulations. As required by the CAA, any facility in these subcategories classified as a major source must implement MACT standards, which are outlined in the proposed regulation. An emission audit or estimate can be performed to determine whether a facility is considered a major source. The equation for estimated emissions, given in the regulations for both subcategories, is dependent on the number of different types of ASTs, number of pumps, and valves, in addition to the facility throughput. An emissions audit involves accurately determining the emissions from sources within a facility. If the emissions determined by either method are greater than 10 tons per year (tpy) or more for any one HAP or 25 tpy or more for any combination of HAPs, the facility is classified as a major source.

As part of this regulation, the EPA has included an analysis for the base year 1998 indicating that there will be approximately 403,600 facilities in the entire gasoline distribution network. Pipeline breakout stations and bulk gasoline terminals comprise a total of 1,300 facilities. Of this total only 20 pipeline breakout stations and 280 bulk gasoline terminals qualify as major sources and require MACT standards.

Once a facility is classified as a major source, the actual MACT standards are straightforward. The gasoline distribution loading rack of a major source facility must limit the emissions to 10 milligrams of total organic compounds per liter of gasoline by the installation of a vapor collection and processing system. ASTs greater than or equal to 20,000 gallons (75 m^3) at a major source facility must have MACT controls specifically referenced as 40 CFR 60.112b(a)(1) through (4), which are the NSPS Kb controls. These MACT controls must be implemented before startup of new facilities and no later than February 8, 1997, for facilities already in operation.

Petroleum Refineries NESHAP

The EPA has proposed NESHAPs (issued on July 15, 1994) for source category of petroleum refineries. In addition to proposing the NESHAPs, the rule will amend two NSPS that currently exist in 40 CFR 60: standards of performance for equipment leak of volatile organic compounds in the synthetic organic chemicals manufacturing industry; and standards of performance for volatile organic compounds emissions from petroleum refinery wastewater systems. The proposed change is different from past regulations in that the EPA is co-proposing two different formats for existing ASTs. Therefore, the proposal does not give much insight into regulation specifics—only the options that the EPA is investigating.

The proposed NESHAP regulations would set MACT standards for the petroleum refineries, which include facilities engaged in producing gasoline, naphtha, kerosene, jet fuels, distillate fuel oils, residual fuel oils, lubricants, or other products made from crude oil or unfinished petroleum derivatives. Determining if a petroleum refinery must conform to this MACT standard would depend on whether the plant site would be considered a major source based on the emissions of HAPs from not only the petroleum refinery process, but the entire site, which may include other processes such as the SOCMI production, discussed previously. If the plant site has the potential to emit greater than 10 tons per year of particular HAP or 25 tons per year of a combination of HAPs, the petroleum process would be subject to these proposed regulations.

The relationship between the HON regulation and the petroleum refinery NESHAP regulation can best be defined by examining the production and use of the gasoline additive MTBE. MTBE is a SOCMI chemical that can be produced at some petroleum refineries. The feedstock for MTBE is a mixed C4, C5 hydrocarbon stream produced in a fluidized catalytic cracking unit, which is subject to the petroleum refineries NESHAP. However, the process used to produce the MTBE from the feedstock is regulated by the HON.

The most relevant part of the petroleum refineries NESHAP regulation pertains to new and existing ASTs. The MACT floor level control is based on the best-performing 12% of sources and was determined to be: Storage vessels with capacity greater than or equal to 47,000 gallons (177 m^3) storing liquids with a vapor pressure greater than or equal to 1.2 psia (8.3 kPa), must control to a level equal to 40 CFR 60 subpart Kb with the exception of fitting requirements for floating roof tanks. The co-proposed alternate involves ASTs greater than or equal to 40,000 gallons (151 m^3) storing liquids with a vapor pressure greater than 0.75 psia (5.2 kPa), which must meet the controls of subpart Kb, including the floating and roof fittings. The regulations will require that existing ASTs, containing liquids with true vapor pressure of 5.0 psia or greater, must meet these requirements within three years of the regulation publication, whereas the HON required that existing regulations be upgraded at the first scheduled maintenance of the ASTs. For new ASTs, the MACT floor would be control of ASTs equal to or greater than 40,000 gallons (151 m^3) with a vapor pressure of the liquid equal to or greater than 0.5 psia (3.4 kPa) and equal to or greater than 20,000 gallons (76 m^3) with a vapor pressure of the liquid equal to or greater than 11.1 psia (77 kPa). The level of control for new tanks should follow the HON AST requirements previously discussed.

Because the petroleum refinery NESHAP regulation is very much in the draft stage, the final regulations (due out in 1995) should be consulted prior to making a final decision on the construction of a new AST or modification of an existing AST.

CHAPTER 5

Additional Federal Regulations

In addition to those already discussed, there are a number of other regulations, administered by various federal agencies, designed to protect the environment from AST spills or failures and to protect the operators and the general public. The Resource Conservation and Recovery Act (RCRA) regulations are administered by the EPA and were published to protect the environment from the improper disposal of hazardous waste. When problems arise from past mismanagement of hazardous waste, the cleanup of inactive and abandoned hazardous waste sites is covered by the RCRA companion law, or the Comprehensive Environmental Response, Compensation, and Liability Act of 1980 (CERCLA), commonly referred to as Superfund. CERCLA also regulates the planning for and response to any release of oil or a hazardous substance that violates federal regulations. The Occupational Safety and Health Administration (OSHA) administers regulations to protect persons from injury associated with ASTs and stored products.

Resource Conservation and Recovery Act Regulations

The EPA administers regulations that protect the environment against the improper disposal of hazardous waste. For an understanding of the regulations and their implications for the construction and operation of ASTs, the purpose and the historical context of the federal laws that mandate the regulations must be examined. Since the end of World War II, industrial growth fueled by consumer demand for new products such as plastics and high-quality paper has led to an increase in disposal, by both industry and consumers, of large amounts of wastes that often are hazardous to the environment and to humans. In 1965, the Solid Waste Disposal Act was developed to encourage environmentally sound methods of waste disposal. Waste-disposal laws were amended in 1970 and again in 1976 by the Resource Conservation and Recovery Act (RCRA). The laws and subsequent EPA regulations protecting the environment from hazardous wastes have been identified with the title of this act. As stated in 42 USC 6901 et seq., the primary goals of RCRA are:

- To protect human health and the environment from the potential hazards of waste disposal.
- To conserve energy and natural resources.
- To reduce the amount of waste generated, including hazardous waste.
- To ensure that waste is managed in an environmentally sound manner.

RCRA was amended in 1980 and again in 1984 by the Hazardous and Solid Waste Amendment (HSWA), which drastically changed the scope of RCRA, largely because of strong citizen concerns and lobbying efforts. Three subtitles of RCRA set forth the framework for the EPA's current regulations on waste management programs:

- *Subtitle C* requires control of hazardous waste from generation to ultimate disposal.

- *Subtitle D* establishes a system for the control of primarily nonhazardous solid waste.
- *Subtitle I* regulates toxic substances and petroleum products stored in underground tanks.

The laws on hazardous waste are of particular interest to AST owners and operators because of the use of ASTs to store and treat hazardous waste. Therefore the focus of this examination of RCRA laws and regulations will be on Subtitle C and its subsequent effect on AST operation. The associated regulations written and administered by the EPA in response to RCRA are found in 40 CFR Subpart I, entitled "Solid Waste," and contained in Parts 260 through 281.

A definition of hazardous waste must be formulated to determine if an AST operator storing a particular waste is governed by the RCRA regulations. A hazardous waste first must be classified as a solid waste, which RCRA defines as garbage, refuse, or sludge that can exist as a solid, semisolid, liquid, or contained gas. A solid waste classified as a hazardous waste must contain, in part or as a whole, a substance that is hazardous based on its characteristics or comes from a specific hazardous waste list. The characteristics of a hazardous waste are as described in 40 CFR 261.20:

- *Ignitability*—Wastes that create fires under certain conditions.
- *Corrosivity*—Wastes such as acids, or those that corrode metal.
- *Reactivity*—Unstable wastes that can create explosions or toxic fumes.
- *Toxicity*—Wastes that can be harmful when ingested or absorbed.

The extensive list published by the EPA is located in 40 CFR 261.30 and is divided into three categories: source-specific waste, generic waste, and commercial chemical products. Generators must evaluate their wastes to determine if any of the four characteristics are exhibited, or if a stored waste is specifically listed in the regulations.

Hazardous Waste Handler Standards

The EPA estimates that 96% of hazardous waste is disposed of on-site, with the other 4%, which still accounts for approximately 12 million tons per year, disposed of off-site. To prevent the improper disposal of hazardous waste into the environment, RCRA has set up a program often referred to as "cradle-to-grave" because it is designed to track the hazardous waste from initial generation to final disposal. The RCRA laws and regulations list specific requirements for three different waste handlers:

- Large and small generators.
- Transporters.
- Treatment, storage, and disposal facilities (TSDFs).

Small generators amass 220 to 2,200 pounds of hazardous waste per month and do not accumulate over 13,200 total pounds, whereas large generators produce over 2,200 pounds per month. Each waste handler must have an EPA identification number to be able to handle hazardous waste. To track the disposal of this waste, the RCRA program requires that the generator complete a one-page uniform manifest document that must accompany the hazardous waste to its final disposal. Each handler of the waste must complete part of the form, with the manifest making a complete circle and returning to the generator within 35 days or the generator must locate the hazardous waste. As part of the Hazardous and Solid Waste Amendment of 1984, generators also must certify on the manifest that they are minimizing the amount and toxicity of hazardous waste, and that the path to final disposal is chosen to reduce the risk to human health and the environment.

RCRA also outlines the transfer of RCRA authority to the states, which have the ultimate enforcement powers of the program. As is the case for all federal regulations, each state has the responsibility to enforce the regulations and has the authority to adopt more stringent regulations to protect the environment within the boundaries of the state. RCRA requires that each state submit to the EPA a RCRA implementation program, which must be at least as stringent as the RCRA regulations. Therefore, it is essential that state and local jurisdictions be consulted for any variances from the federal program.

To store hazardous waste according to RCRA regulations, waste handlers must obtain a permit. Waste handlers required to obtain permits are:

- Owners and operators of TSDFs.
- Small generators that store waste for more than 180 days (or 270 days if the generator must ship the waste over 200 miles).
- Large generators that accumulate waste for more than 90 days.
- Transporters storing hazardous waste for more than 10 days.

Because there are approximately 4000 hazardous waste TSDFs and 50 states had to become qualified by the EPA to administer the program, Congress knew that the EPA would need a considerable amount of time to approve the operating permits. Therefore, RCRA contains two separate regulations, termed "permitted" and "interim," given in 40 CFR 264 and 265, respectively. To qualify for interim status, TSDFs had to be in existence prior to November 19, 1980, and must operate under 40 CFR 265. In order to continue to operate, TSDFs must have received a final permit before October 1992. New TSDFs cannot obtain interim status and must obtain a permit before the facility becomes operational.

ASTs are widely used at TSDFs, so subsequent discussion of RCRA will focus on this phase of hazardous waste handling. The TSDFs are required to meet certain general standards that are nearly identical for both the interim and permitted facilities; they must:

- Obtain an EPA identification number to operate.
- Analyze and identify waste prior to treatment, storage, or disposal.
- Comply with the RCRA program manifest procedure and other reporting and recordkeeping procedures.
- Prevent the entry of unauthorized personnel into a facility by installing fences and surveillance systems.
- Periodically inspect the facility for any problems.
- Adequately train personnel to operate the facility and respond to a leak or a spill.
- Prepare a contingency plan and establish procedures for an emergency. SPCC Plans can be modified to incorporate the RCRA program.

The RCRA regulations also list specific standards for the hazardous waste treatment, storage, and disposal phases. These regulations address standards for the various components of a facility, which may include containers, tank systems, surface impoundments, waste piles, landfills, and incinerators, and also operational standards that include proper management, financial responsibility, closure of facility, and emission standards. Subpart J outlines the standards for tanks systems of the respective permitted (40 CFR 264) and interim status (40 CFR 265) TSDFs, which are nearly identical with variances noted in subsequent discussion. Subpart J divides the standards into sections under the following categories:

- Assessment of existing tank system integrity.
- Design and installation of new tank systems or components.
- Secondary containment and leak detection.
- General operating requirements.
- Tank system and component inspections.
- Leak or spill response.
- Closure and post-closure care.

Assessment of existing ASTs that do not have secondary containment must include owner or operator determination that the tank system is not leaking and/or unfit for use. Such inspection was to be reviewed and certified by a professional engineer attesting to tank integrity by January 12, 1988. The assessment also was to determine whether the tank system was of adequate integrity and compatible with the contents to ensure that the tank system would not collapse, rupture, or fail. At a minimum, the assessment must also consider the design standard used for the system, hazardous characteristics of the waste, existing corrosion protection, documented age of the tank, and the result of leak tests. For existing tanks changing service after July 14, 1986, to store hazardous waste, the assessment must be completed within 12 months.

New tank assessment is similar to existing tank assessment and also must be certified by a professional engineer. A corrosion expert must assess the cathodic protection requirements for any metal that is in contact with the earth or water. The owner or the operator is obligated to install the system recommended by the corrosion expert to guarantee protection. The assessment also should examine the foundation to ensure that the design will carry the load of a full tank and resist natural loads such as wind, seismic, and frost heave. The owner must ensure proper handling of the tank system during installation and conduct a tightness test to verify that there are no leaks. Prior to placing a new tank into service, a professional engineer or inspector must certify that the installation is free of structural or protective system defects.

The most controversial part of the RCRA regulations regarding tank systems is the requirement of secondary containment and leak detection. A tank's secondary system must be capable of detecting, collecting, and containing releases until the waste can be safely removed. To meet these requirements, secondary containment systems minimally should be constructed of, or lined with, a material that is compatible with the hazardous waste being stored and is designed to detect leaks in both the secondary and the primary containment within 24 hours after the release. The form of secondary containment shall be a liner, a vault, a double-walled tank, or an approved equal, and should be part of a new tank system or added to existing tanks over a period of time. The time frame for implementation of secondary containment depends on the age of the tank, and is based on an EPA study finding that leaks in ASTs are more likely to occur in older tanks, with the useful life of a tank determined to be 15 years. Therefore, for tanks of a known age, secondary containment was to be added within two years of January 12, 1987, or at 15 years of age, which ever occurred later. If the age of the tank is unknown, it should be assumed to be seven years or to be equal to the age of the facility. An owner can obtain a variance on secondary containment for existing tanks only, but must demonstrate that the secondary containment is not necessary.

The regulations also list standards for the operation and the inspection of an AST system and the response to a leak. The system should have adequate devices, such as check valves and overflow protection, to prevent spills. The system also must be inspected daily, especially tank components that do not have secondary containment. In response to an AST system leak of greater than one pound, the owner or the operator must take corrective action to prevent filling of the system, to remove the spill, and to empty the tank within 24 hours. The AST system must be repaired and assessed by a professional engineer or an inspector before it can be put back into service.

CERCLA Regulations

Growth in new products over the years has led to the development and increased use of hazardous substances and subsequently an increased environmental threat. This has resulted in laws and regulations designed to inform the general public of the hazardous substances stored or released from a given facility, and the cleanup of hazardous substances that have been released or illegally disposed. Two laws were enacted by Congress to handle the increased presence of hazardous substances: the Comprehensive Environmental Response, Compensation, and Liability Act (CERCLA) (42 USC 9605 et seq.) in 1980 and the Superfund Amendments and Reauthorization Act (SARA) (Pub. L. 99-499) in 1986. The law, the associated amendments, and the regulations are hereinafter known as CERCLA. In addition to amending CERCLA, SARA contained a number of free-standing amendments which included Title III, known as the "Emergency Planning and Community Right-to-Know Act of 1986," which has four provisions: emergency planning, emergency release notification, community right-to-know reporting, and toxic chemical release inventory. The CERCLA regulations are given in Subpart J of 40 CFR, entitled "Superfund, Emergency Planning and Community Right-to-Know Programs," and are the focus of discussion in this section.

Inventory Reporting

The community right-to-know and the toxic chemical release provisions of Title III were designed to increase the public's knowledge and access to information on the presence of hazardous chemicals in their communities and the release of these chemicals into the environment. This act requires that certain facilities report key information to the local emergency planning committee (LEPC) and the general public regarding certain chemicals produced or stored at a facility. For hazardous chemicals, the reporting is in the form of both the material safety data sheets (MSDSs) and the emergency and hazardous chemical inventory form. A facility must retain the MSDS forms, which are produced by the chemical manufacturer and provide general information on all hazardous chemicals. The facility is required by Title III to submit the MSDS forms or a list of the MSDS chemicals to the state emergency response commission (SERC), the LEPC, and local fire departments. The EPA has established chemical quantities that a facility must report, equal to 500 pounds or the threshold planning quantity for extremely hazardous substances and 10,000 pounds for hazardous chemicals. The same facilities also are required to submit an annual emergency and hazardous chemical inventory form, which uses a two-tier approach. Tier I requires information such as the maximum amount of hazardous chemicals at the facility during the preceding year, an estimate of the average daily amount of hazardous chemicals at the facility, and the general location of the chemicals. Tier II information is submitted only upon the request of the emergency entity and includes, in addition to the Tier I information, specific chemical information, a brief description of the specific location, and the manner of storage. A toxic chemical release form must be filed, stating the routine emissions of 300 specific chemicals from facilities of SIC codes 20 through 39. The purpose is to inform the general public and government officials about routine releases into the environment.

Emergency Release Notification

The emergency notification provisions and the subsequent regulations come into play when a hazardous substance, in the form of a liquid, a solid, or a gas, is released into the environment (air, water, or land). The reporting of a hazardous substance release is divided into two regulation sections, depending on the hazard potential. Section 40 CFR 302 outlines the procedure for reporting a release of hazardous substances as required by the Comprehensive Environmental Response, Compensation, and Liability Act of 1980 (Pub. L. 96-510). These substances, commonly referred to as the CERCLA hazardous substances, are tabulated in 40 CFR 302.4, along with the reportable quantity (RQ) and the threshold planning quantity (TPQ). SARA Title III expanded the reporting of hazardous materials by adding a list of extremely hazardous substances (EHS), found in Section 40 CFR 355 of the EPA regulations. Petroleum products are specifically exempt from the CERCLA hazardous substance list, but if the petroleum product contains an EHS, such as benzene, in excess of 1%, the facility must become part of the planning and reporting requirements of SARA Title III.

An owner or an operator in charge of a facility must become part of the planning and reporting requirements of the CERCLA regulations when the storage of hazardous substances is in excess of the TPQ limits. This calculation also must take into account the amount of an extremely hazardous substance present in mixtures or solutions in excess of 1%. The TPQs are based on a quantity that can cause significant harm to the environment upon release and were intended to assist local governments in directing their planning efforts.

Emergency Planning

To minimize the environmental impact associated with a release of oil or hazardous substances, both CERCLA and the OPA of 1990 require response planning. The planning of a discharge was set up under CERCLA and expanded under the OPA of 1990 to include release of oil. The planning began with the formation of the National Response Team (NRT), by Executive Order 12777 in 1991, and the National Contingency Plan (NCP), which provides efficient, coordi-

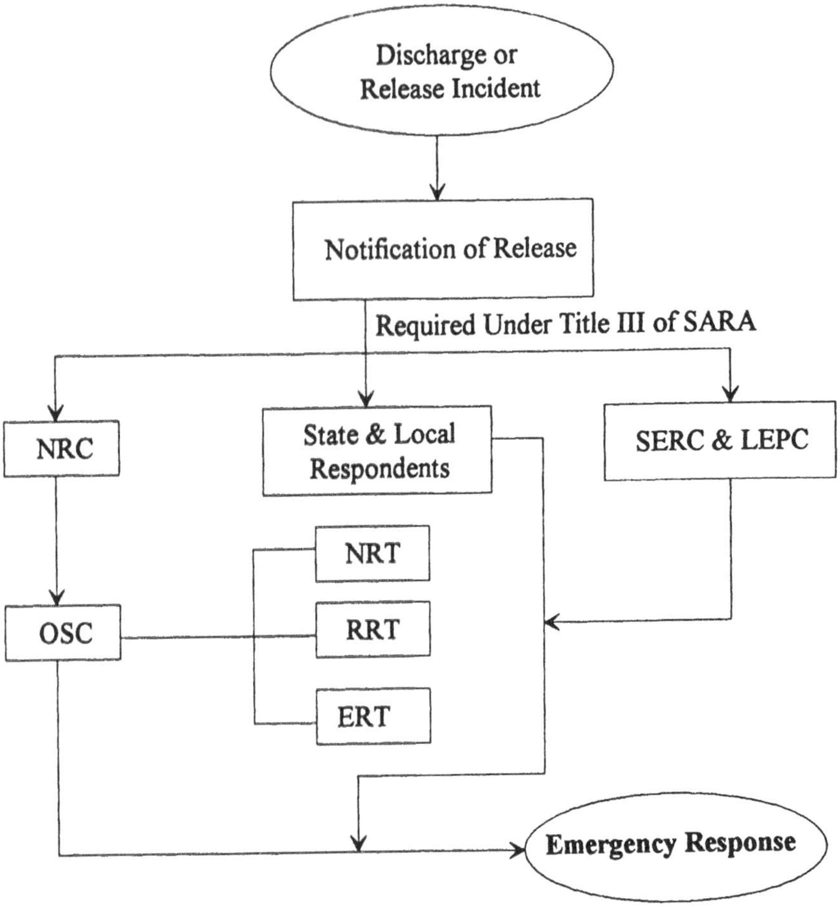

FIGURE 5-1.
National response system.

safety and health of employees. Part 29 of the USC contains the labor regulations written by OSHA to accomplish the task of protecting employees. This section examines three subparts that directly affect employees who operate or work on ASTs, but which represent only a fraction of the regulations.

Subpart D of 29 CFR 1910, Walking-Working Surfaces, outlines regulations regarding the design of, access to, and movement around an AST. These regulations detail the design of handrails (§1910.23) used to protect against a fall from a stairway or a platform greater than four feet in height. Per OSHA regulations, handrails are to be 42 inches high with top and intermediate rails constructed of 2″ × 2″ × 3/8″ angle iron or 1½″ pipe and a 4″ × 1/4″ toeboard positioned 1/4 inch off the walking surface. The handrail height for stairways should be between 30 and 34 inches from the top of the handrail to a line along the front edge of the stair tread. The actual design of stairways (§1910.24) also is given in the OSHA regulations, along with that of fixed ladders (§1910.27). Because OSHA does not address the design of spiral stairways for tanks, the requirements outlined in API 650 should be used, which are based on OSHA regulations but contain adaptations for tanks. The width of the spiral stairways should be 24 inches with the treads 8 inches wide and constructed of grating or a nonskid material. The angle of the stairways should be less than 50 degrees, with API 650 recommending that the sum of twice the rise plus the run be between 24 and 26 inches. According to OSHA, fixed ladders shall be equipped with 3/4 inch minimum diameter rungs spaced at one foot with the distance between the side rails equal to 16 inches. Straight ladders without cages should ascend less than 20 feet. Above 20 feet, a safety device or cage must be installed with specific details provided by OSHA. An intermediate landing must be provided with a maximum ladder ascent of 30 feet or 20 feet, for a caged ladder or a ladder with a safety device, respectively. The regulations also provide minimum clearances from all sides of the ladder to guarantee that obstructions will not interfere with the climber.

An event that has increased in frequency over the past few years is entry into existing tanks for inspection or repair purposes. Section 29 CFR 1910.146 outlines entrance into a confined space that may or may not require a permit. The regulations define a confined space as one that is large enough for an employee to enter it and perform an assigned task but has limited or restricted means of entry and is not designed for continuous occupancy. Because ASTs fit the description of a confined space, they must be examined to determine the need for an entry permit. For a confined space to require a permit, or controlled and supervised entry, the space must have at least one of the following characteristics:

- It must contain or have the potential to contain a hazardous atmosphere.
- It must contain a material that has the potential for engulfing an entrant.

- It must have an internal configuration such that an entrant could be trapped or asphyxiated by inwardly converging sides of the space.
- It must contain another recognized serious safety or health hazard.

If the confined space requires a permit for entry, one should consult the regulations to understand the extensive requirements for safe entry. The regulations do contain a flowchart in the appendix to assist an owner, operator, or other party in deciphering the confined space entry regulations.

The OSHA regulations (Subpart H, §1910.101 et seq.) also control the design of ASTs containing hazardous material. For simplicity, the discussion here focuses on the storage of flammable and combustible liquids (§1910.106), which are defined by the product flash point, a format identical to the classification of NFPA 30 (Table 6-1). The OSHA regulations outline the standards to follow for the tank and piping design and venting of the tank, as well as its installation.

The OSHA regulations generally require that ASTs be designed by using steel and a nationally recognized standard such as API 650, UL 142, or ASME Section VIII. Construction and testing also should be in accordance with the applicable standard and adhered to per OSHA. The regulations also detail venting of ASTs that have various design pressures, located inside or outside a building. The venting should be checked by utilizing API 2000 with a tank having either a frangible joint or emergency venting to prevent tank failure during a fire. A tank installed inside a building must rely not on a frangible roof joint but, rather, an emergency vent designed to release vapors to the outside. Discussion of frangible roof joint and emergency venting is reserved for Chapters 9 and 11.

The regulations also outline the location of a tank with respect to other tanks and the design of diked areas. The spacing of tanks is based on tank diameter and their design with respect to venting, with a minimum of three feet required. Tanks shall be installed with either a diked wall around the tanks or a drainage system to relocate spilled product to a remote impounding area. The diked or impound area should be designed to hold the largest tank capacity, with the regulations also restricting the diked wall height to a minimum of three feet for an earthen wall and six feet for dike walls.

CHAPTER 6

National Fire Protection Association Codes

Fire codes are widely used guidelines for the safe storage of flammable and combustible liquids in ASTs. The National Fire Protection Association, a nationally recognized organization, publishes codes that are often referred to in industry standards and are adopted by state fire marshall officials in their local regulations. The most relevant AST-related NFPA codes are discussed in this chapter. These codes include: NFPA 30, *Flammable and Combustible Liquids Code;* NFPA 30A, *Automotive and Marine Service Station Code;* and NFPA 11, *Low Expansion Foam and Combined Agent Systems.*

National Fire Protection Association, Code 30

The National Fire Protection Association issued NFPA 30, *Flammable and Combustible Liquids Code*, 1993 Edition, with an effective date of August 1993. This code is recommended for use as a guideline to fire officials and other "authorities having jurisdiction" to reduce potential hazards in the use of flammable and combustible liquids. This code is not a regulation, but it may be incorporated into a legal regulation by a city, a municipality, a state, or the federal government.

NFPA covers both shop-built and field-erected ASTs in this document. This code is divided into six chapters, but for the focus of this book, only the applicable sections of NFPA 30 are highlighted.

General Provisions of NFPA 30

NFPA 30 shall apply to all flammable and combustible liquids, including waste liquids, with the exception of liquids that are solid at 100°F or above and liquefied gases or cryogenic liquids. The goal of NFPA 30 is to list requirements for the safe storage of flammable and combustible liquids. Because of the variety of these readily available liquids, a classification system is needed to differentiate them according to characteristics. Refer to Table 6-1 for the appropriate category designations.

Table 6-1. Flammable and combustible liquids

Flammable Liquid: A liquid having a closed cup flash point below 100°F and having a vapor pressure not exceeding 40 psia at 100°F, which shall be known as a Class I liquid.	
Class IA	shall include those liquids having flash points below 73°F and a boiling point below 100°F.
Class IB	shall include those liquids having flash points below 73°F and a boiling point at or above 100°F.
Class IC	shall include those liquids having flash points at or above 73°F and below 100°F.
Combustible Liquid: A liquid having a closed cup flash point at or above 100°F.	
Class II	shall include those liquids having flash points at or above 100°F and below 140°F.
Class IIIA	shall include those liquids having flash points at or above 140°F and below 200°F.
Class IIIB	shall include those liquids having flash points at or above 200°F.

Source: Developed from NFPA 30.

When storing flammable and combustible liquids, one should keep in mind that the volatility of these liquids will be increased upon exposure to a heat source. For example, when Class II or Class III liquids are subjected to a particular storage or use condition where the liquids would be excessively heated to above their flash points, additional safeguards might be necessary. In this case, appropriate exposure, venting, or diking requirements should be explored. Also, one must be aware of liquids that possess unusual burning characteristics or are highly reactive with other substances. It may be necessary to implement additional safe storage requirements above those for normal liquids of similar flash points.

NFPA makes a strong effort to allow this code to be flexible and user-friendly by allowing the authority having jurisdiction, at its discretion, to alter the code requirements on the basis of special features or considerations. NFPA recognizes that a fire code must be integrated into the local community according to the community's ability to handle potential fire hazard situations. The authority having jurisdiction may grant a variance for a specific part of the NFPA 30 code. For example, the capacity and the construction of the proposed ASTs combined with the characteristics of the liquid to be stored offer a degree of safety above the intent of the code; so the authority having jurisdiction will allow reduced distance requirements from the nearest property line. As the intent of this code is fire protection, it is feasible that other regulations (i.e., environmental protection) will impose requirements not listed by this code. NFPA 30 states that alternative arrangements affording protection that is at least equivalent to that required by this code shall be permitted to be used.

Tank Storage

Atmospheric tanks (tanks designed to operate at pressures from atmospheric to 0.5 psig) shall be built to recognized standards of design or approved equivalents as shown in Table 6-2. Tanks shall be fabricated out of metal and shall be welded, riveted and caulked, or bolted in a manner consistent with sound engineering design. Low-pressure tanks and pressure vessels shall be permitted to be used as atmospheric tanks, but a liquid must not be stored at a temperature at or above its boiling point in an atmospheric tank.

For low-pressure tanks (tanks designed to withstand an internal pressure between 0.5 psig and 15.0 psig) and pressure vessels (tanks

Table 6-2. Atmospheric AST codes of design

Underwriters Laboratories Inc., *Standard for Steel Aboveground Tanks for flammable and Combustible Liquids,* UL 142
American Petroleum Institute Standard No. 650, *Welded Steel Tanks for Oil Storage,* Eighth Edition
American Petroleum Institute Specification 12B, *Bolted Tanks for Storage of Production Liquids,* Twelfth Edition
American Petroleum Institute Specification 12D, *Field Welded Tanks for Storage of Production Liquids,* Eighth Edition
American Petroleum Institute Specification 12F, *Shop Welded Tanks for Storage of Production Liquids,* Seventh Edition

Source: Developed from NFPA 30.

designed to withstand an internal pressure greater than 15.0 psig), the normal operating pressure of the tank shall not exceed the design pressure of the tank. The recognized standards of design for low-pressure tanks are:

- American Petroleum Institute Standard No. 620, *Recommended Rules for the Design and Construction of Large, Welded, Low-Pressure Storage Tanks,* Fifth Edition.
- The principles of the *Code for Unfired Pressure Vessels,* Section VIII, Division I of the ASME *Boiler and Pressure Vessel Code.*

For pressure vessels, the standards of design are:

- ASME *Boiler and Pressure Vessel Code,* Section I (Power Boilers).
- ASME *Boiler and Pressure Vessel Code,* Section VIII, Division I or II (Pressure Vessels).

In addition, pressure vessels may be operated as low-pressure tanks per NFPA 30. An important exception for low-pressure tanks refers to Underwriters Laboratories Standard 142. UL 142 tanks shall be permitted to be used for operating pressures not exceeding 1.0 psig and shall be limited to 2.5 psig under emergency venting conditions. Table 6-3 summarizes the differences between atmospheric tanks, low-pressure tanks, and pressure vessels.

Location of ASTs

In this section, the location of aboveground tanks is highlighted. NFPA 30 sets its distance requirements for ASTs by the type of liquid being stored. Tables 6-4 to 6-8 will list the NFPA 30 safe distance recommendations for stable liquids (operating pressures above and

Table 6-3. Tank type, operating range, AST design codes

Type	Operating Range	AST design codes
Atmospheric tanks	atmospheric–0.5 psig	UL 142, API 650, API 620, API 12D, API 12F
Low-pressure tanks	0.5 psig–15.0 psig	API 620, ASME Section VIII—Division I
Pressure vessels	above 15.0 psig	ASME Section I, ASME Section VIII—Division I or II

Source: Developed from NFPA 30.

below 2.5 psig), boil-over liquids, unstable liquids, and Class IIIB liquids. Table 6-9 should be used as a supplemental safe distance reference table for Tables 6-4, 6-5, and 6-7. NFPA 30 does make one recommendation for pressure vessels in addition to the distance requirements that should be noted: "Where end failure of horizontal pressure tanks and vessels can expose property, the tank should be placed with the longitudinal axis parallel to the nearest important exposure." In other words, do not place the end of a horizontal pressure vessel facing any important building, property line, or public access, because of potential danger of an end failure.

Table 6-4 lists distance requirements for ASTs storing Class I liquids, Class II liquids, Class IIIA liquids, unstable liquids, or liquids with boil-over characteristics operating at pressures not in excess of 2.5 psig and designed with a weak shell-to-roof seam or equipped with emergency venting devices that will not permit pressures to exceed 2.5 psig. For the purpose of this section, a floating roof tank is defined as:

1. A pontoon or double-deck metal floating roof in an open-top tank in accordance with API Standard 650, or
2. A fixed metal roof with ventilation at the top and roof eaves in accordance with API Standard 650 and containing a metal floating roof or cover meeting any of the following requirements:
   (a) A pontoon or double-deck metal floating roof meeting the requirements of API Standard 650.
   (b) A metal floating cover supported by liquid-tight metal floating devices that provide sufficient buoyancy to prevent the liquid surface from being exposed when half of the flotation is lost.

As a point of clarification, an internal metal floating roof that does not meet the requirements of item (2) above, or one that uses plastic foam (except for seals) for flotation, even if encapsulated in metal or fiberglass, shall be considered a fixed roof tank. One exception should be noted for vertical tanks. If the vertical tanks have a weak

Table 6-4. Stable liquids (operating pressure of 2.5 psig or less)

Type of Tank	Protection	Minimum distance in feet from property line that can be built upon (not less than 5.0 ft)	Minimum distance in feet from nearest side of any public way or important building on the same property (not less than 5.0 ft)
Floating roof	Protection for exposures*	1/2 times diameter of tank	1/6 times diameter of tank
	None	Diameter of tank, but need not exceed 175 ft	1/6 times diameter of tank
Vertical with weak roof-to-shell seam	Approved foam or inerting system on tanks not exceeding 150 ft dia.	1/2 times diameter of tank	1/6 times diameter of tank
	Protection for exposures*	Diameter of tank	1/3 times diameter of tank
	None	2 times diameter of tank, but need not exceed 350 ft	1/3 times diameter of tank
Horizontal and vertical with emergency relief venting at 2.5 psig	Approved inerting system on the tank or approved foam system on vertical tanks	1/2 times Table 6-9	1/2 times Table 6-9
	Protection for exposures*	Table 6-9	Table 6-9
	None	2 times Table 6-9	Table 6-9

Source: NFPA 30.
*Protection for exposures shall be defined as fire protection for structures on property adjacent to liquid storage.

shell-to-roof seam and are storing Class IIIA liquids, then the tanks may be located at one-half the distances specified in Table 6-4, provided that the tanks are not within a diked area or drainage path for tank storing a Class I or Class II liquid.

Table 6-5 lists distance requirements for ASTs storing Class I liquids, Class II liquids, or Class IIIA liquids, except those liquids with boil-over characteristics and unstable liquids, operating at pressures exceeding 2.5 psig or equipped with emergency venting devices that will permit pressures to exceed 2.5 psig.

Table 6-5. Stable liquids (operating pressures greater than 2.5 psig)

Type of tank	Protection	Minimum distance in feet from property line that can be built upon	Minimum distance in feet from nearest side of any public way or important building on the same property
Any type	Protection for exposure*	1/2 times Table 6-9, but not less than 25 ft	1½ times Table 6-9, but not less than 25 ft
	None	3 times Table 6-9, but not less than 50 ft	1½ times Table 6-9, but not less than 25 ft

Source: NFPA 30.
*Protection for exposures shall be defined as fire protection for structures on property adjacent to liquid storage.

Table 6-6 lists distance requirements for ASTs storing liquids with boil-over characteristics. Liquids with boil-over characteristics shall not be stored in fixed roof tanks larger than 150 feet in diameter unless an approved inerting system is provided on the tank.

Table 6-6. Boil-over liquids

Type of Tank	Protection	Minimum distance in feet from property line that can be built upon (not less than 5.0 ft)	Minimum distance in feet from nearest side of any public way or important building on the same property (not less than 5.0 ft)
Floating roof	Protection for exposures*	1/2 times diameter of tank	1/6 times diameter of tank
	None	Diameter of tank	1/6 times diameter of tank
Fixed roof	Approved foam or inerting system	Diameter of tank	1/3 times diameter of tank
	Protection for exposures*	2 times diameter of tank	2/3 times diameter of tank
	None	4 times diameter of tank, but not to exceed 350 ft	2/3 times diameter of tank

Source: NFPA 30.
*Protection for exposures shall be defined as fire protection for structures on property adjacent to liquid storage.

Table 6-7 lists distance requirements for ASTs storing unstable liquids.

Table 6-7. Unstable liquids

Horizontal/vertical tanks with emergency relief venting to:	Protection	Minimum distance in feet from property line that can be built upon	Minimum distance in feet from nearest side of any public way or important building on the same property
Permit pressure not in excess of 2.5 psig	Protected tank**	Table 6-9, but not less than 25 ft	Not less than 25 ft
	Protection for exposures*	1/2 times Table 6-9, but not less than 50 ft	Not less than 50 ft
	None	5 times Table 6-9, but not less than 100 ft	Not less than 100 ft
Permit pressure over 2.5 psig	Protected tank**	2 times Table 6-9, but not less than 50 ft	Not less than 50 ft
	Protection for exposures*	4 times Table 6-9, but not less than 100 ft	Not less than 100 ft
	None	8 times Table 6-9, but not less than 150 ft	Not less than 150 ft

Source: NFPA 30.
*Protection for exposures shall be defined as fire protection for structures on property adjacent to liquid storage.
**Protected tank is defined as a tank protected with any one of the following: approved water spray, approved inerting, approved insulation and refrigeration, or approved barricade.

Table 6-8 lists distance requirements for ASTs storing Class IIIB liquids, excluding unstable liquids. An exception is made for a Class IIIB AST located within a diked area or drainage path for a tank(s) storing Class I or Class II liquid. In this situation, Table 6-4 shall apply or, if the tanks are vertical and have a weak shell-to-roof seam, then the tanks may be located at one-half the distances specified in Table 6-4.

Table 6-9, as noted above, is a reference table for Tables 6-4, 6-5, and 6-7.

Table 6-8. Class IIIB Liquids

Tank capacity in gallons	Minimum distance in feet from property line that can be built upon	Minimum distance in feet from nearest side of any public way or important building on the same property
12,000 or less	5	5
12,001 to 30,000	10	5
30,001 to 50,000	10	10
50,001 to 100,000	15	10
100,001 or more	15	15

Source: NFPA 30

Table 6-9. Reference table for use in Tables 6-4, 6-5, and 6-7

Tank capacity in gallons	Minimum distance in feet from property line that can be built upon	Minimum distance in feet from nearest side of any public way or important building on the same property
275 or less	5	5
276 to 750	10	5
751 to 12,000	15	5
12,001 to 30,000	20	5
30,001 to 50,000	30	10
50,001 to 100,000	50	15
100,001 to 500,000	80	25
500,001 to 1,000,000	100	35
1,000,001 to 2,000,000	135	45
2,000,001 to 3,000,000	165	55
3,000,001 or more	175	60

Source: NFPA 30.

Spacing between ASTs

This section describes safe spacing distances (shell-to-shell) between any two adjacent ASTs. Table 6-10 lists minimum tank spacing requirements for ASTs storing Class I, Class II, or Class IIIA stable liquids. For tanks storing unstable liquids, the separation distances between such ASTs shall not be less than one-half the sum of their diameters. (*Example:* Tank #1 has a diameter of 100 feet, and Tank #2 has a diameter of 80 feet; so, at a minimum these tanks must be separated by 90 feet.) For tanks in a diked area containing Class I or Class II liquids, or in the drainage path or Class I or Class II liquids, that are compacted in three or more rows or in an irregular pattern, special requirements may be imposed. The authority having jurisdiction can assess greater spacing guidelines or other means to make tanks in the interior of the pattern accessible to firefighting personnel.

The minimum separation distance between a liquid propane (LP)-gas container and a Class I, Class II, or Class IIIA liquid storage tank shall be 20 feet. An exception will be granted in the case of a Class I, Class II, or Class IIIA liquid storage tank operating at pressures exceeding 2.5 psig or equipped with emergency venting that will permit pressures to exceed 2.5 psig. In this case, Table 6-10 and Exception A (of Table 6-10) shall apply. Appropriate measures should be taken to prevent accumulation of Class I, Class II, or Class IIIA liquids underneath adjacent LP-gas containers by using dikes, diversion curbs, or grading. If flammable or combustible ASTs are located within a diked area, the LP-gas containers shall be located outside of the diked area at least 10 feet away from the centerline of the wall of the diked area. These provisions shall not apply to LP-gas containers of 125 gallons or less capacity that are installed adjacent to fuel oil tanks of 660 gallons or less capacity.

Control of Spillage from ASTs

In today's environment, controlling spillage from ASTs is a major issue. NFPA 30 states that facilities shall be provided with remote impounding or tank diking so that any accidental discharge of any Class I, Class II, or Class IIIA liquids will be prevented from endangering important facilities and adjoining property or reaching

Table 6-10. Minimum tank spacing (shell-to-shell)

Tank Size (Note Exception A & B)	Floating Roof Tanks	Fixed or Horizontal Tanks Storing Class I or II Liquids	Fixed or Horiizontal Tanks Storing Class IIIA Liquids
All tanks not over 150 feet in diameter	1/6 sum of adjacent tank diameters but not less than 3 ft	1/6 sum of adjacent tank diameters but not less than 3 ft	1/6 sum of adjacent tank diameters but not less than 3 ft
Tanks larger than 150 feet in diameter;			
If remote impounding[1]	1/6 sum of adjacent tank diameters	1/4 sum of adjacent tank diameters	1/6 sum of adjacent tank diameters
If impounding by means of diking[2]	1/4 sum of adjacent tank diameters	1/3 sum of adjacent tank diameters	1/4 sum of adjacent tank diameters

Source: Developed from NFPA 30.

(1) Remote impounding is accomplished by draining any liquid to a remote impounding area away from any other tanks so that impounded liquid will not be held against any tank.

(2) Impounding by means of diking is accomplished by building dike walls around tanks to protect surrounding property and waterways.

Exception A: Crude petroleum tanks having individual capacities not exceeding 126,000 gallons (3,000 barrels), where located at production facilities in isolated locations, need not be separated by more than 3 feet.

Exception B: Tanks used only for storing class IIIB liquids shall be permitted to be spaced no less than 3 feet apart unless within a diked area or drainage path for a tank storing a Class I or II liquid, in which case the provision of Table 6-10 shall apply.

waterways. Class IIIB liquids are exempted from this provision and do not require any special drainage or diking provisions for fire protection purposes. In addition, aboveground secondary containment type tanks (i.e., double wall tanks built to UL 142 standard) need not meet the above requirements if all of the following conditions are met:

1. Tank capacity shall not exceed 12,000 gallons.
2. All piping connections shall be above normal maximum liquid level.
3. Means are provided to prevent release of liquid from the tank by siphon flow.
4. Means shall be provided for determining the liquid level in the tank.
5. Operators install an overfilling alarm that sounds at 90% of tank capacity, as well as an automatic shutoff valve to the delivery truck that will shut off at 95% of tank capacity.
6. There is a minimum space of 3 feet between adjacent tanks.

7. Collision barriers or fencing is provided around tanks to protect the tanks from damage by a motor vehicle.
8. Where the interstitial space is enclosed (double wall tank), emergency venting of the tank shall be provided.

For tanks that do not have attached secondary containment, the protection of adjoining property or waterways can be obtained by using remote impounding or by diking around the tanks.

Remote impounding is accomplished by means of drainage of the accidental spill to a remote impounding area so that the impounded liquid will not be held against tanks. To use this system, the following conditions shall apply:

1. There shall be a minimum 1% slope away from the tank for at least 50 feet from the impounding area.
2. The impounding area shall have a capacity not less than that of the largest tank that can drain into it. If there is not enough open area around the tanks for 100% capacity remote impounding, a combination of remote impounding and tank diking can be employed.
3. The route of the drainage system shall be located so that if the liquids in the drainage system are ignited, the fire will not seriously expose tanks or adjoining property.
4. The impounding area must be a minimum of 50 feet from any tank or property line.

Another method to control spillage from ASTs is by impounding around tanks by means of diking. To use this system, the following provisions shall apply:

1. There shall be a minimum 1% slope away from the tank for at least 50 feet or to the dike base, whichever is less.
2. The volumetric capacity of the diked area shall not be less than the greatest amount of liquid that can be released from the largest full tank within the diked area. To allow for the volume occupied by tanks, the capacity of the diked area enclosing more than one tank shall be calculated after deducting the volume of the tanks, other than the largest tank, below the height of the dike.
3. The outside base of the dike at ground level must be a minimum of 10 feet away from any property line.

4. The walls of the diked area shall be constructed of earth, steel, concrete, or solid masonry designed to be liquid-tight and to withstand a full hydrostatic head. Earthen walls 3 feet or more in height should have a flat section on top that is a minimum of 2 feet wide. For storing Class I liquids in a diked area with porous soil conditions, special treatment may be necessary to prevent seepage into waterways during an accidental spill.
5. Dikes shall be restricted to an average interior height of 6 feet above interior grade, except as provided in item (6) below.
6. Dikes may be higher than 6 feet if provisions are made for normal access and emergency access to tanks, valves and other equipment while allowing for safe passage from the diked area. Additionally:
   (a) Where the average height of the dike containing Class I liquids is over 12 feet above the interior grade or where the distance between any tank and the top inside edge of the dike wall is less than the height of the dike wall, provisions should be made for normal operation of valves and for access to tank roofs without entry into the area below the top of the dike. Options would include elevated walkways, remote-operated valves, or similar arrangements.
   (b) Piping passing through the dike wall must be designed to withstand excessive stresses as a result of settlement or fire exposure.
   (c) The minimum distance between tanks and the bottom of the dike wall shall be 5 feet.
7. In order to prevent spills from endangering adjacent tanks within the diked area, each diked area containing two or more tanks shall be subdivided, preferably by drainage channels or at least by intermediate dikes, as follows:
   (a) For storing normally stable liquids in vertical cone roof tanks or floating roof tanks or for storing crude petroleum in producing areas in any type of tank, one subdivision should be made for each tank in excess of 10,000 barrels (bbl) or 420,000 gallons (ga), and one subdivision should be made for each group of tanks (no tank exceeding 10,000 bbl capacity) having an aggregate capacity not exceeding 15,000 bbl or 630,000 ga.
   (b) For storing normally stable liquids in tanks not covered in item (7a), provide one subdivision for each tank in excess of 2,380 bbl or 100,000 ga, and one subdivision for each group

of tanks (no tank exceeding 2,380 bbl capacity) having an aggregate capacity not exceeding 3,570 bbl or 150,000 ga.

(c) For storing unstable liquids in any type of tank, provide one subdivision for each tank. *Exception:* Tanks installed in accordance with the drainage requirements of NFPA 13, *Standard for Water Sprayed Fixed Systems for Fire Protection,* shall require no additional subdivision. *Note:* Because unstable liquids will react more rapidly when heated than when at ambient temperatures, subdivision by drainage channels is the preferred method.

(d) Whenever two or more tanks storing Class I liquids, any of which is over 150 feet in diameter, are located in a common diked area, intermediate dikes shall be provided between adjacent tanks to hold at least 10% of the capacity of the tank so enclosed, not including the volume displaced by the tank.

(e) The drainage channels or intermediate dikes shall be located between tanks so as to take full advantage of the available space with due regard for the individual tank capacities. Intermediate dikes, where used, shall not be less than 18 inches in height.

8. Where provision is made for draining water from diked areas, such drains will be controlled in such a manner as to prevent flammable or combustible liquids from entering public sewers or drains if their presence would constitute a hazard. Control of drainage shall be accessible under fire conditions from outside the dike.
9. Storage of combustible materials, empty or full drums, or barrels shall not be allowed within the diked area.

Accessories/Openings for Aboveground Tanks

Normal Venting for Aboveground Tanks

Atmospheric storage tanks, low-pressure tanks, and pressure vessels shall be adequately vented to prevent the development of vacuum or pressure, as a result of filing or emptying, and atmospheric temperature changes, from exceeding the design pressure of the tank or the vessel.

Tanks and pressure vessels storing Class IA, Class IB, and Class IC liquids shall be equipped with venting devices that shall be normally closed except when venting to pressure or vacuum conditions. (Alternatively, tanks storing Class IB and Class IC liquids may be vented with listed flame arrestors.) Tanks of 3,000 bbl/126,000 ga capacity or less containing crude petroleum in crude-producing areas, and outside aboveground atmospheric tanks under 23.8 bbl/1,000 ga capacity containing other than Class IA liquids, shall be permitted to have open vents.

Emergency Relief Venting for Aboveground Tanks

Every aboveground storage tank shall have some form of construction or device that will relieve excessive internal pressure caused by exposure fires. *Exception:* Tanks larger than 285 bbl/12,000 ga capacity storing Class IIIB liquids and not within the diked area or drainage path of Class I or Class II liquids do not require emergency relief venting.

For vertical tanks, the "form of construction" referred to above may take the form of a floating roof tank or a weak roof-to-shell seam. The weak roof-to-shell seam shall be constructed to fail before any other seam does (refer to API 650 or UL 142 for design methods). Pressure relief devices include a self-closing manhole cover, a long-bolt manway that permits the cover to lift under internal pressure, or an additional or larger relief value or valves.

Vent Piping for Aboveground Tanks

Where vent piping outlets for tanks storing Class I liquids are adjacent to buildings or public ways, they shall be located so that vapors are released at a safe point outside of buildings and not less than 12 feet above the adjacent ground level. Vent pipes should be at least five feet from building openings. The manifolding of vent piping shall be avoided except where required for special purposes such as vapor recovery, vapor conservation, or air pollution control. Operators must not manifold vent piping of Class I liquids with vent piping for tanks storing Class II or Class III liquids unless positive means are provided to prevent vapor contamination.

Tank Openings (Other Than Vents) for Aboveground Tanks

An internal or an external valve shall be attached to each connection on an AST through which liquid can normally flow. For each connection below liquid level where liquid does not normally flow, the connection should be a liquid-tight closure (valve, plug, or blind). Openings for gauges on tanks storing Class I liquids shall be provided with a vapor-tight cover that shall be kept closed when not gauging the tank. Filling and emptying connections for Class I, Class II, or Class III liquids shall be located outside buildings at a location free from ignition sources and not less than five feet from any building opening. Such connection points shall be closed and liquid-tight when not in use (e.g., by a remote, lockable steel fill box with a nonsparking aluminum cover that houses the filling connection).

Supports, Foundations, and Anchorage for All Tank Locations

Tanks shall rest on the ground or on foundations made of concrete, steel, or masonry. Where tanks are supported above the foundations, the tank supports shall be installed on firm foundations. The supports shall be constructed of steel, concrete, or masonry. If steel supports, they should be protected by materials having a two-hour fire rating. (*Exception:* Steel saddles need not be protected if less than 12 inches high at their lowest point. For horizontal, shop-built ASTs most manufacturers build their saddles to be less than 12 inches to avoid this requirement.) All tanks should be evenly supported to prevent excessive concentration of loads on any part of the shell. In areas subject to flooding, provisions shall be made for tank anchorage to a foundation to prevent the tanks from floating during a rise in water level. In areas subject to earthquakes, tank supports shall be designed to resist damage to such shock.

Prevention of Overfilling Tanks

ASTs at terminals receiving Class I liquids shall prevent overfilling of tanks by using one of the following methods:

- High liquid level detection devices that will alert personnel to shut down the liquid flow.
- High liquid level detection devices that can automatically shut down the liquid flow.
- Personnel monitoring the flow who can relay information to other personnel who can shut the flow down.

Piping Systems

All piping systems should be installed with proper support and connections according to manufacturers' recommendations. Valves should perform their intended function and not fail under pool fire conditions. Also, numerous values should be installed so that the tank may be isolated in case of catastrophic failure or fire. Proper draining away from pipe or pipe racks shall be used to eliminate liquid pooling underneath the pipes.

Testing and Maintenance

Before a tank is put into service, it should be tested according to the testing requirements of the applicable code under which the tank was built (i.e., UL, ASME, API). This applies to both field-erected and shop-built tanks. Testing methods vary according to the type of tank in question, but, in general, hydrostatic testing is employed for field-erected tanks and air pressure for shop-built tanks (a double wall shop-built tank may use a vacuum to test the interstitial space). Refer to Chapters 9 and 16 for further information on tank testing. All tanks should be thoroughly inspected and any damage fixed before they are put into service. Regular maintenance will ensure that a tank remains liquid-tight.

National Fire Protection Association, Code 30A

The National Fire Protection Association issued NFPA 30A, *Automotive and Marine Service Station Code,* 1993 Edition, with an effective date of August 1993. This code is recommended for use as a guideline to fire officials and other "authorities having jurisdiction" to reduce potential hazards at service stations. This code is not a regulation, but it may be incorporated into a legal regulation by a city, a municipality, a state, or the federal government. NFPA 30A covers shop-built ASTs in this document, which is divided into nine chapters. The sections of this code that relate to marine service stations are not covered in this text.

General Provisions of NFPA 30A

The purpose of this code is to recommend provisions for the safe storage of flammable and combustible liquids at automotive and marine service stations, but it does not apply to natural gases being used as automotive fuels. Storage of liquids shall be in listed AST systems, and accurate daily inventory records should be kept or leak detection monitoring should be employed for any indication of tank

Table 6-11. Distance requirements for ASTs at automotive service stations

Tanks shall be located at least:
(a) 50 feet from the nearest important building on the same property.
(b) 50 feet from any fuel dispenser.
(c) 50 feet from the nearest side of any public way.
(d) 100 feet from any property line that is or might be built upon, including the opposite side of a public way.

Source: Developed from NFPA 30A.

Exception #1: All distances shall be permitted to be reduced by 50% if the tanks are fire-resistant tanks or are installed in vaults.

Exception #2: At commercial, industrial, governmental, or manufacturing establishments, where the tanks are intended for fueling corporate vehicles (business fleet), no minimum distance shall be required if the tanks are fire-resistant tanks or are installed in vaults.

or piping failure. ASTs at bulk plants should not dispense their contents directly into customer fuel tanks unless the bulk plant operations are made to be distinct from the retail fueling area. This can be accomplished by installing a fence or a wall to separate the two areas. Also, ASTs at bulk plants should not be piped directly to service station tanks.

NFPA 30A allows for installation of tanks inside buildings if the NFPA 30 distance requirements cannot be met, and the authority having jurisdiction approves the special installation. These type of buildings are referred to as "special enclosures" and they shall be constructed of six inches of reinforced concrete with only top openings for inspection. Special enclosures should be liquid- and vapor-tight, should require no backfill, and must possess a means to vent the enclosed space at all times.

ASTs at Automotive Service Stations

Refer to Table 6-11 for distance requirements of tanks storing Class I and Class II liquids at an individual site. ASTs at service stations shall be limited to an individual capacity of 12,000 gallons and an aggregate capacity of 40,000 gallons. It is important to note that the authority having jurisdiction may increase or decrease the size and the quantity of tanks on the basis of site-specific details. Control of Spillage from ASTs shall be in conjunction with the requirements of NFPA 30 as previously listed. Figure 6-1 illustrates NFPA 30A distance requirements for UL 142 unprotected ASTs, while Figure 6-2 shows the requirements for fire-resistant or vaulted tanks with dispensers mounted on top of the ASTs.

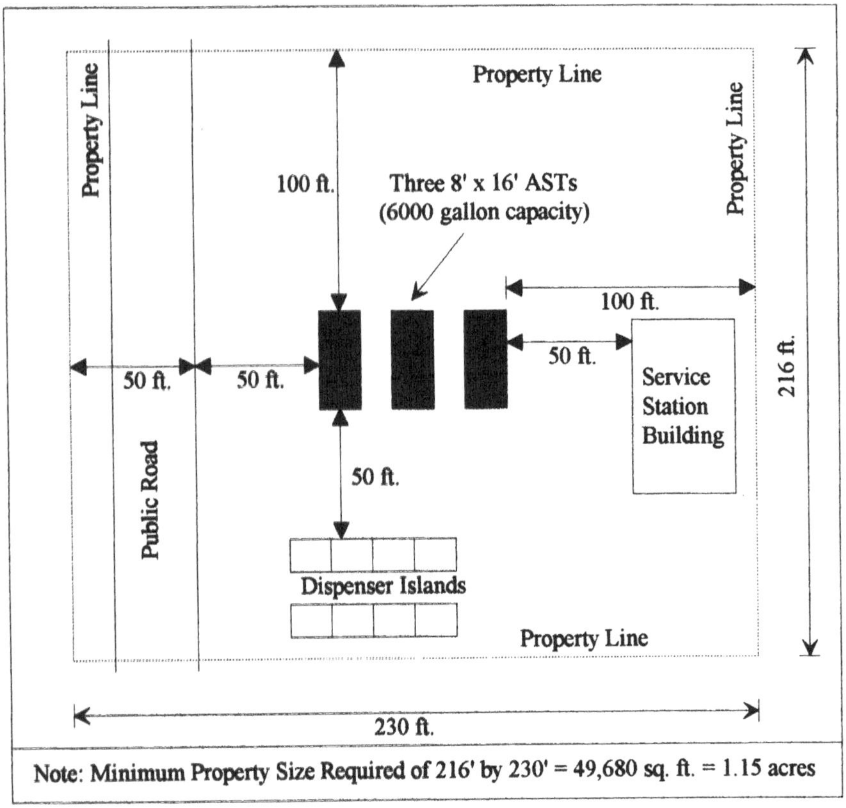

FIGURE 6-1.
Distance requiurements for ASTs at automotive service stations per NFPA 30A. This figure depicts UL142 tanks without fire protection.

Fire-Resistant Tanks

In defining a fire-resistant tank, NFPA 30A states that the construction of the tank providing the fire protection capabilities shall "prevent release of liquid, failure of the primary tank, failure of the supporting structure, and impairment of venting for a period of not less than two hours." NFPA 30A refers to UL 2085, *Outline of Investigation for Insulated Aboveground Tanks for Flammable and Combustible Liquids,* or its equivalent as the testing standard for a fire-resistant tank. A fire-resistant tank only can have required openings on the top of the tank and should be appropriately anchored. Also, vehicle collision barriers should be installed to protect the tank.

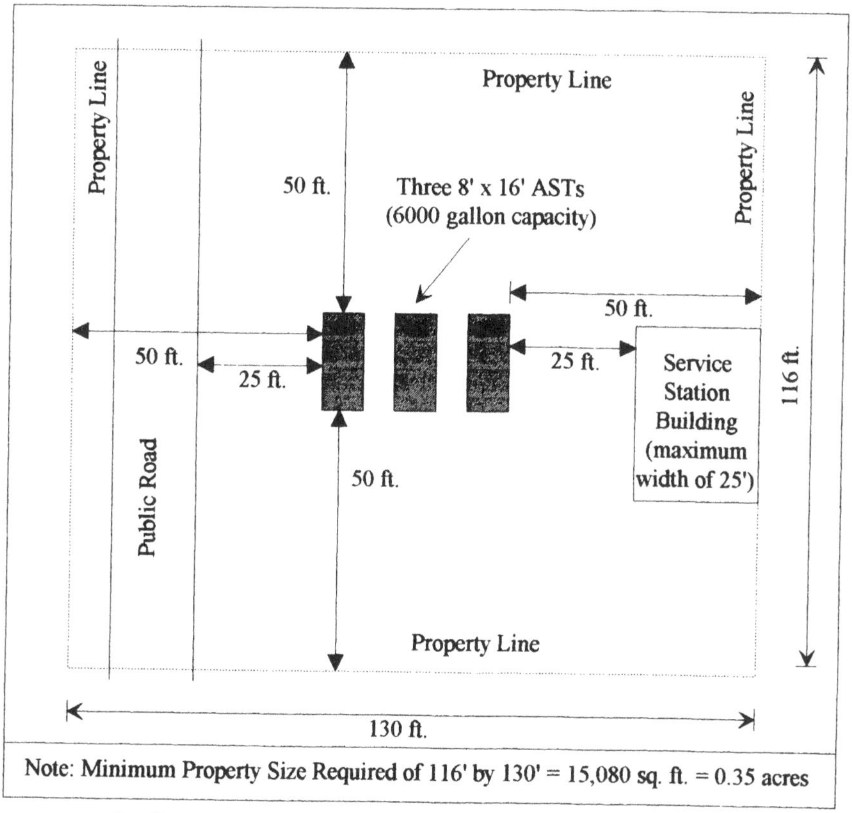

FIGURE 6-2.
Distance requiurements for ASTs at automotive service stations per NFPA 30A. This figure depicts fire resistant or vaulted tanks with dispensers mounted on top of the ASTs.

Vaults

Vaults provide a means of avoiding UST regulations while at the same time offering a degree of security similar to that of a UST installation. NFPA 30A calls for the walls and the floor of the vault to be made of six inches of reinforced concrete at a minimum, but the top shall be made of a noncombustible material that will be weaker than the walls so as to direct an explosion inside the vault upward. A vault may be either above or below grade, and only a listed UL 142 AST can be installed. (*Note:* Do not use a UL 58 UST because it is not constructed as a self-supporting structure.) Each tank should have its own vault, but adjacent vaults may share a common wall. Backfill should not be put around the tank, and there should be enough space surrounding the tank for operators to visually inspect the tank. An entry point should be provided with each vault for personnel, but individuals should not enter the vault until all the vapors have been purged by means of venting. NFPA 30A states that each vault should be equipped with a leak detection system that will activate an alarm, should the presence of a liquid be detected.

Tank-Filling Operations

A liquid level monitor shall be installed in the AST that will sound an audible alarm when the tank reaches 90% of capacity. Also, an automatic shutoff valve should be utilized during filling operations to stop the flow of liquid into the tank when the liquid reaches 98% of capacity, or a flow-restriction valve shall be used that will restrict the flow of liquid to a maximum of 2.5 gallons per minute when the liquid in the tank reaches 95% of capacity. The delivery vehicle shall be separated from any aboveground tank by at least 25 feet (tanks that are filled by gravity do not require any separation distance), and the delivery operator should be aware of the available tank capacity. ASTs should be filled through a liquid-tight connection utilizing either a quick-connect or dry-break coupling.

Physical Protection

For tanks not in vaults, a six-foot-high chain link fence should be installed around them. The fence should be at least ten feet from the closest tank and have a lockable gate. In addition, suitable vehicular barriers must be installed to protect the tanks from collision.

Fuel-Dispensing System

Class I and Class II liquids shall be suction-pumped from the tanks to a listed dispenser unit. The dispenser shall have a control that permits the dispenser to operate only when the dispensing nozzle is removed from its bracket or normal position, and the control is manually activated. The dispensing device should be mounted and bolted on a concrete island or its equivalent. The hose attached to the dispenser shall be listed and is not to exceed 18 feet at an automotive service station.

National Fire Protection Association, Code 11

Another pertinent AST code developed by the National Fire Protection Association is NFPA 11, *Low Expansion Foam,* 1994 Edition. NFPA 11 outlines various types of foam systems to be used for tank fire protection. The use of foam is one of the most popular methods of preventing fire from spreading to tanks, and it is employed extensively to extinguish storage tank fires. A lighter substance than water, foam is widely used because it forms a floating blanket over flammable or combustible liquids that prevents or extinguishes fire by eliminating air and cooling the liquids.

For fixed roof storage tanks, protection of flammable liquids can be accomplished by installing foam discharge outlets equally spaced around the circumference of the AST. The discharge outlets should be high enough on the shell so that the tank contents do not flow into the foam lines. A fixed foam system or portable foam application equipment may be used in conjunction with the discharge outlets. Table 6-12 lists appropriate numbers of discharge outlets, based on the size of the vertical fixed roof atmospheric tank.

For fixed roof storage tanks, the code also allows for a subsurface application. For ASTs storing hydrocarbons with a flash point below 100°F, the number of discharge outlets is the same as given in Table 6-12. The discharge outlets are in the shell, near the bottom, and discharge foam into the product. This foam then rises to the surface of the product.

Open-top floating roof tanks are tanks without fixed roofs that have double deck or pontoon-type floating roofs. For floating roof tanks, there are two common designs for foam application from stationary outlets: top-of-seal application and below primary seal applications. The top-of-seal application is used to provide foam discharge above a mechanical shoe seal, a tube seal weathershield, or a secondary seal (Figure 6-3). The below primary seal application used to provide foam discharge below a mechanical shoe seal, a metal weathershield, or a metal secondary shield (Figure 6-4). Both applications require the use of a foam dam, which is a 10 gauge thick steel plate welded to the floating roof around the circumference of the AST. The purpose of the foam dam is to retain foam around the seal area by covering the seal with foam and allowing the foam to

Table 6-12. Fixed roof tank foam discharge outlet quantity

Tank diameter (feet)	Minimum number of discharge outlets
Up to 80	1
81–120	2
121–140	3
141–160	4
161–180	5
181–200	6
Over 201	one additional for each 5000 sq ft

Source: Developed from NFPA 11.

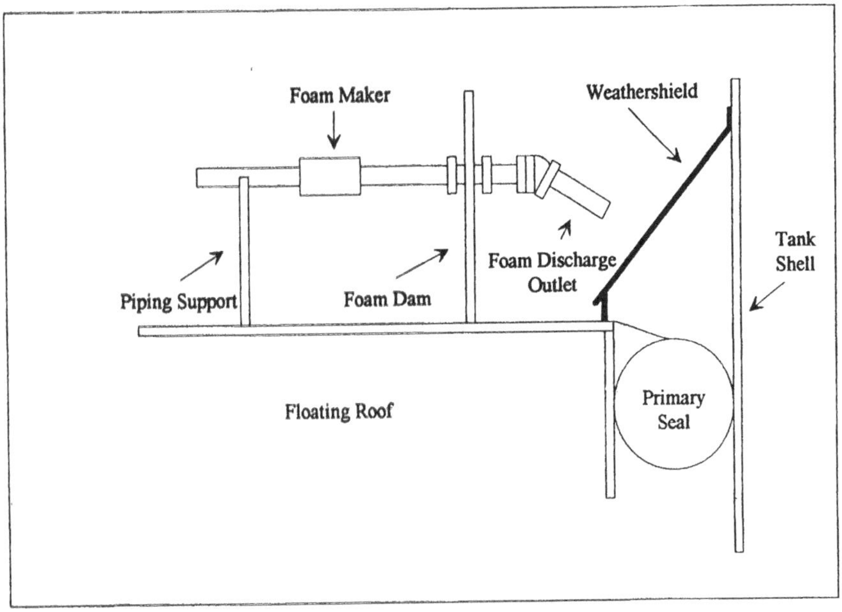

FIGURE 6-3.
Top-of-seal foam application.

flow to the point of the seal rupture. Refer to Figure 6-5 for an example of a double seal floating roof utilizing a foam dam.

Outdoor covered (internal) floating roof tanks are defined within this standard as vertical cylindrical tanks with a fixed metal roof—cone or geodesic dome—equipped with ventilation at the top and containing a metal double deck or pontoon-type floating roof or

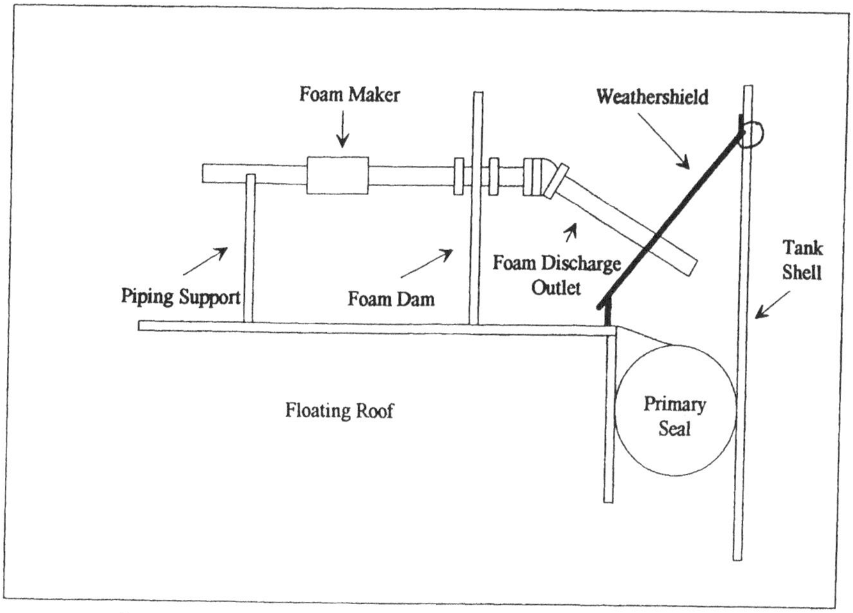

FIGURE 6-4.
Bellow primary seal foam application.

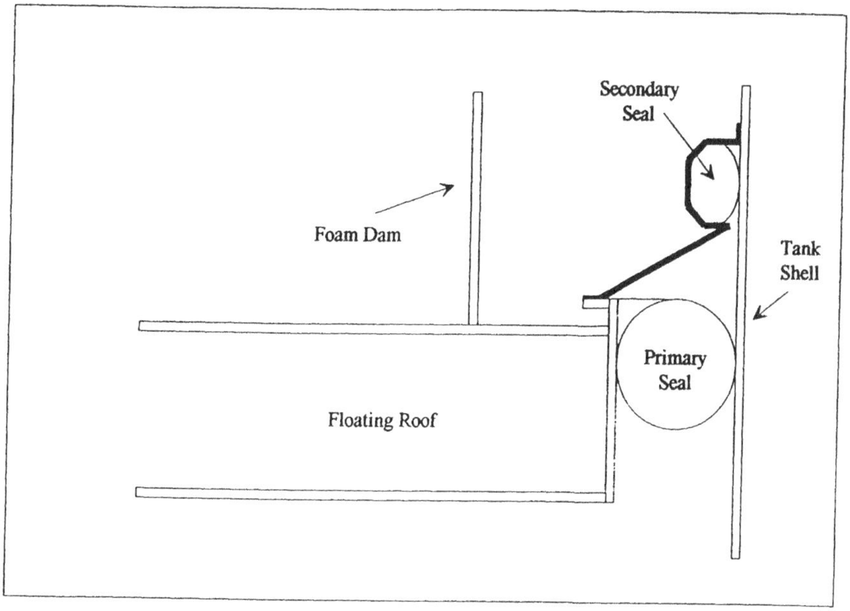

FIGURE 6-5.
Double seal floating roof utilizing a foam dam.

metal floating cover. Internal floating roof tanks shall have fire systems designed for a full surface fire, much like a fixed cone roof storage tank, or designed for a seal fire, by installing a foam dam, in the same manner as an external floating roof. For other types of internal floating roof tanks, protection of flammable liquids is to be accomplished in the same manner as for a fixed cone roof storage tank.

In addition, NFPA 11 suggests that the AST owner submit plans and specifications to the authority having jurisdiction for approval of the fire protection system. In conclusion, this standard has sections addressing: installation requirements of foam application equipment, testing and acceptance of installed foam systems, and maintenance of foam systems.

CHAPTER 7

Additional Fire Codes

In addition to the National Fire Protection Association (NFPA) codes, three other fire codes should be discussed. The focus of the fire codes, as outlined in this chapter, mainly is on shop-built ASTs. Of the three codes, the Uniform Fire Code (UFC) is the one most often referred to by fire officials, and it is widely adopted in the western United States. UFC has two sections relating to ASTs, UFC Appendix II-F and UFC Appendix Standard A-II-F-1. The Building & Official Code Administrators (BOCA) publish subsection F-2805.5 dealing with ASTs. BOCA is most popular in the northeastern section of the United States. The Southern Building Code Congress International (SBCCI) publishes the Standard Fire Prevention Code, which addresses ASTs. SBCCI is most prevalent in the southeastern United States.

The similarity between these fire codes is the use of fire-resistant ASTs to dispense fuel into motor vehicles, and these codes sometimes are specified in unison. For example, in efforts to get an AST installation approved by the authority having jurisdiction, it is conceivable that an official will require that the AST installation meet both the NFPA 30A code and the BOCA code.

Uniform Fire Code, Appendix II-F

UFC Appendix II-F, *Protected Aboveground Tanks for Motor Vehicle Fuel Dispensing Stations Outside Buildings*, was approved by the International Fire Codes Institute (IFCI) and was first included in the 1991 edition of the UFC,* and since then has been revised for the 1994 edition. Appendix II-F deals with storage in, and dispensing of motor fuels into the fuel tanks of motor vehicles from, protected aboveground tanks located outside of buildings. For the purposes of Appendix II-F, UFC defines a protected aboveground tank as "a listed tank system consisting of a primary tank provided with protection from physical damage, and fire-resistive protection from a high-intensity liquid pool fire exposure. The tank system is allowed to provide these protection elements as a unit or is allowed to be an assembly of components, or a combination thereof."

The first step in the AST installation process, according to the UFC, is to apply for a permit from the authority having jurisdiction. An installation plan must be submitted in conjunction with the permit application. That plan must include, but is not limited to, the design details of the installation as highlighted in Table 7-1.

Tank Design/Capacity

UFC Appendix II-F dictates that all tanks be listed (by a third party such as UL) protected aboveground tanks meeting the requirements of UFC Standard A-II-F-1. In addition, primary tanks have a maximum individual capacity of 10,000 gallons and a maximum aggregate capacity of 40,000 gallons at a facility.

*Reproduced from the 1991 edition of the *Uniform Fire Code*™, copyright © 1991, with the permission of the joint publishers, the Western Fire Chiefs Association and the International Conference of Building Officials.

Table 7-1. UFC installation plan details

Installation application items
• Quantities and types of liquids to be stored.
• Distances from tanks and dispensers to property lines and buildings.
• Vehicle access.
• Fire appliances.
• Vehicle impact protection.
• Protected aboveground tanks and their supports.
• Method of storage and dispensing.
• Overfill prevention, spill containment, vents, vapor recovery, dispensers, and other equipment.
• Seismic design in accordance with the building code.
• Secondary containment.
• Venting.
• Piping.
• Electrical systems.
• Emergency controls.
• Any other information as required by the fire chief.

Source: Developed from UFC Appendix II-F.

Tank Installation

A primary concern in the installation of ASTs is separation distance requirements. UFC Appendix II-F, as in NFPA, set forth safe separation distances between protected aboveground tanks and property lines, buildings, and other tanks. Refer to Table 7-2 for minimum separation distances for protected ASTs, while Figure 7-1 depicts these distance requirements.

Table 7-2. UFC Appendix II-F separation distances

Individual tank capacity	Minimum distance from property line which is or can be built upon, including the opposite side of a public way	Minimum distance from the nearest side of any public way or from the nearest important building on the same property	Minimum distance between tanks
≤ 6,000 gallons	15 ft	5 ft	3 ft
> 6,000 gallons	50 ft	25 ft	3 ft

Source: UFC Appendix II-F.

One exception should be noted for separation distances. If the tank installation has the maximum allowable aggregate capacity of

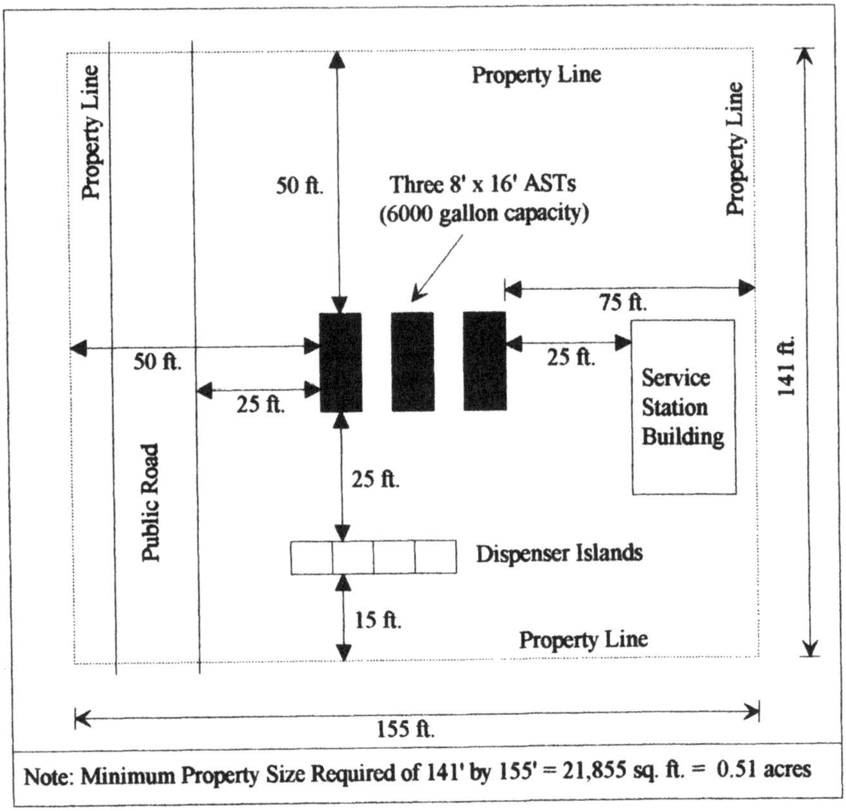

FIGURE 7-1.
Distance requirements for ASTs of automotive service stations per UFC. This figure depicts fire-resistant tanks or vaults with UL 142 tanks.

40,000 gallons, then the installation must be separated from other protected AST installations by 100 feet. UFC Appendix II-F requires that all protected AST installations have secondary containment that is monitored either visually or automatically. The secondary containment can be accomplished in either of two ways:

1. By integral secondary containment that is part of the listed protected tank system.
2. By means of diking or drainage control (remote impounding) in accordance with specific provisions of the UFC.

Integral (enclosed) secondary containment systems shall be supplied with emergency venting, and vehicle impact protection shall be provided for each protected AST and associated piping subject to vehicle impact by means of guardposts or other approved barrier protection. Openings for protected ASTs shall be through the top only. Consequently, any fittings or drains below liquid level are prohibited. Protected ASTs are allowed to have dispensers installed on top of or immediately adjacent to the tank.

Tank Filling

As with NFPA 30 and 30A, overfill preventon is required by UFC Appendix II-F. Also, Appendix II-F requires that an overfill prevention system be provided for each protected aboveground tank and that the tank not be filled in excess of 90% of its capacity. Table 7-3 highlights UFC Appendix II-F requirements for the overfill prevention system during tank-filling operations.

Table 7-3. UFC Appendix II-F overfill prevention system requirements

During tank-filling operations, UFC requires:
• Tank-filling personnel must be notified when the liquid level has reached 85% of capacity by an audible or visual alarm, a tank level gauge that indicates capacity, or an equivalent means of protection.
• A means to automatically shut off the flow of fuel to the tank at 90% of tank capacity.
• Appropriate signage/charting that will enable the tank-filling personnel to determine the appropriate gallonage to fill the tank to 90% of capacity before attempting to fill.
• Each fill connection/fitting shall be provided with an appropriate five-gallon-minimum overspill container that can manually drain into the primary tank. For remote filling connections, a remote or a portable spill container is required.
• Warning signs to be posted around the protected ASTs listing the potential hazards and prohibiting simultaneous tank filing and fuel dispensing.

Source: Developed from UFC Appendix II-F.

Uniform Fire Code, Appendix Standard A-II-F-1

UFC Appendix Standard A-II-F-1, *Testing Requirements for Protected Motor Vehicle Fuel Storage Tanks*, was approved by the International Fire Codes Institute (IFCI) and was first included in the 1994 edition of the code. The scope of this standard is to test protected aboveground tanks used for storing and dispensing flammable or combustible liquid motor fuels into motor vehicles as mandated by UFC Appendix II-F.

Most tank manufacturers want to have an entire multi-capacity protected AST line approved in one test; hence a representative tank size to be tested was established as the tank with the "greatest ratio of fire exposed surface to tank volume." By using this method, the unit tested does not necessarily have the smallest tank capacity, and the test will be representative of the manufacturer's entire line of ASTs. Other conditions prior to testing include:

- Installation of all normal and emergency venting.
- The ambient air temperature at the beginning of the test must be within the range of 50–90°F.
- All tank materials are to be in normal operating condition.
- All nonferrous external tank accessories that are to be installed in the tank openings shall be installed for the test to determine the potential for such accessories to melt and allow an explosive condition to occur.

Type of Test

The Uniform Fire Code Appendix Standard A-II-F-1 requires a two-hour full-scale fire test, which can be either a pool fire or a furnace test as long as the heat flux is maintained. The temperature of the environment that maintains the heat flux shall be between 1800°F and 2200°F after the first five minutes of the test. Appendix Standard A-II-F-1 states: "The test specimen shall be exposed to heat flux and temperature conditions representative of total engulfment in the luminous flame regime of a large free-burning liquid hydrocarbon pool fire." During the test, the tank to be tested should be completely empty. The approved tests should meet the following guidelines:

- *Furnace test*: The furnace shall be capable of maintaining a total heat flux of 65,000 Btu per exposed square feet per hour (±5,000 Btu).
- *Pool fire test*: The average total cold wall heat flux of 50,000 Btu per exposed square feet per hour (±2,500 Btu) averaged over the duration of the test on all exposed surfaces shall be maintained during the two-hour test.
- *Both tests*: The placement of thermocouples is the same for both. Thermocouples are to be placed evenly on the primary tank

with a maximum of nine square feet between each thermocouple, with at least one thermocouple placed on each tank segment and face or end. Thermocouples should be a minimum of 12 inches from any tank opening. Thermocouple readings should be recorded at five-minute intervals for the duration of the two-hour test.

Test Results

A protected AST shall be considered to have successfully passed the fire test if the following conditions have been met:

- The temperature readings from the thermocouples on the primary tank must not exceed an average maximum rise of 260°F, and no individual thermocouple can exceed 400°F during the test.
- The primary tank and its structural supports must maintain their integrity.
- Emergency vents must remain operational at all times.

Additional Testing

Once the fire exposure test has been successfully completed, several additional tests must be conducted: a hose stream test, a leakage test, an impact protection test, a projectile resistance test (when it is required by the authority having jurisdiction), and a climatic/environmental test. The tank system will have passed these tests if, upon their completion, the primary tank remains leak-tight.

Immediately after the fire test, the protected tank shall be subjected to a hose stream test in accordance with the Uniform Building Code Standard 7-1, Section 7.108, *Testing of Wall Assembles*, and then a leakage test must be completed. The leakage test will consist of pressure-testing the primary tank according to the manufacturer's listing procedure. (*Note*: See Chapter 16 for a description of pressure-testing of tanks.) An optional leak test may be performed by completely filling the primary tank with water and maintaining that liquid level for a 24-hour period.

The remaining two tests may be performed on either the tested tank or another tank. If the protected tank is going to be installed without the benefit of guard posts, then the tank must be able to withstand a heavy vehicle impact of 12,000 pounds at ten miles per hour applied over a one square foot area at a height of 18 inches above the bottom of the protected AST.

Projectile protection for ASTs is debatable. For the purpose of UFC Appendix Standard A-II-F-1, if the authority having jurisdiction requires a projectile protection test, then the protected tank shall be tested in accordance with the requirements for bullet resistance according to UFC Section 7702.3.4.3.

Finally, Appendix Standard A-II-F-1 requires that a climatic/environmental test to be conducted on the protected tank to evaluate the tank's ability to perform as required by this standard in climates and environments that a protected tank could be exposed to during its anticipated service life. For example, a protected tank will be exposed to different climatic/environmental characteristics in St. Petersburg, Florida, as compared to Minneapolis, Minnesota.

A protected AST that meets the performance criteria of UFC Appendix Standard A-II-F-1 shall be labeled in the manufacturer's facility prior to installation. The label will indicate compliance with UFC Appendix Standard A-II-F-1 and must be permanently attached to the exterior of the tank in a location that will be visible after the tank is installed in the field.

The Building & Official Code Administrators National Fire Prevention Code

The Building & Official Code Administrators (BOCA) *National Fire Prevention Code*, 1993 Edition, is used by fire officials mostly in the northeastern quadrant of the United States in reference to ASTs. The BOCA language basically states that only protected ASTs are allowed to store motor fuels at facilities generally not accessible to the public (i.e., private fleet fueling locations). BOCA's definition of a protected tank mentions NFPA 30A, a listed and approved tank assembly, or a special enclosure. Therefore, the tank system must have passed either UL 2085, *Outline for Investigation for Insulated Aboveground Tanks for Flammable and Combustible Liquids*, or UFC Appendix Standard A-II-F-1, *Testing Requirements for Protected Motor Vehicle Fuel Storage Tanks*, or must be installed in a special enclosure per NFPA 30A. Table 7-4 highlights additional requirements of the BOCA code.

Figure 7-2 depicts distance requirements for ASTs per the BOCA code.

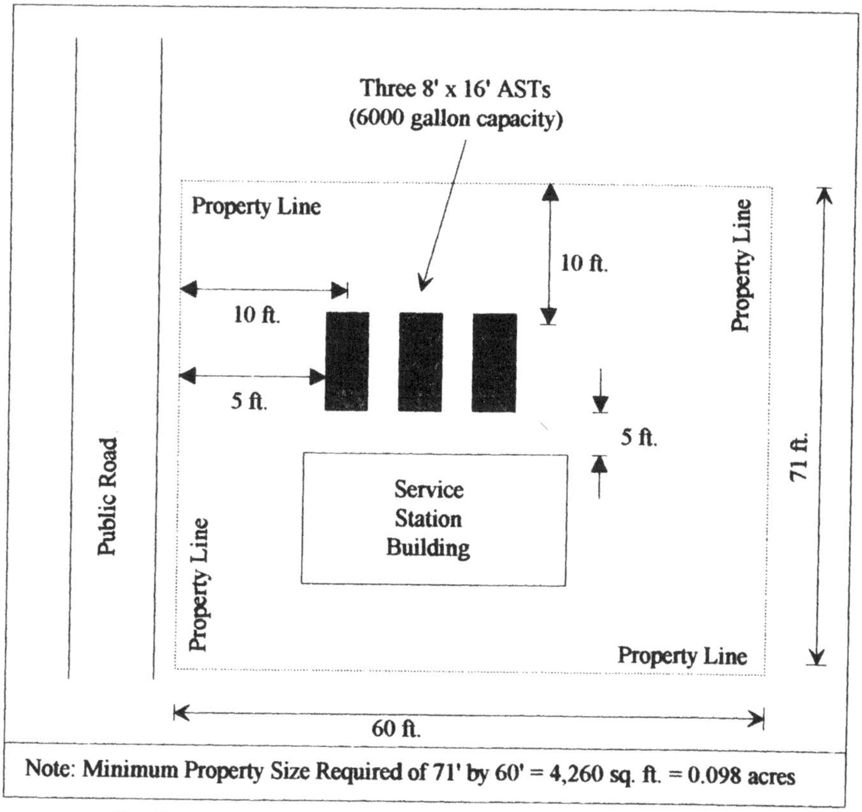

FIGURE 7-2.
Distance requirements for ASTs at automotive service stations per BOCA. This figure depicts fire-resistant ASTs with top-mounted dispensers.

Table 7-4. BOCA Code requirements

The following requirements must be met per BOCA Code:
• Three feet of accessible space surrounding each tank.
• Automatic fuel shutoff devices at 90% of tank capacity.
• All openings and connections to be only through the top of the tank.
• Separation distance from buildings, property lines, and public ways not less than 5 ft.
• Fuel delivery to tanks only allowed by fixed pumps, and anti-siphon devices required.
• Capacity limits: 6,000 gallons individual or 18,000 gallons aggregate capacity.
• Fuel dispensing systems installed per NFPA 30 or 30A.
• Dispensing devices allowed to be mounted directly on top of the tank.
• No minimum separation distance of delivery vehicles from tank.
• Separation distance of 100 ft between 18,000-gallon aggregate capacity tank clusters.
• Physical barrier protection for tanks (guard posts placed within 3 ft of AST spaced 4 ft on centers and minimum 30 inches in height to withstand 12,000-pound force).

Source: Developed from BOCA Code.

As can be seen, BOCA closely follows UFC Appendix II-F and NFPA 30A. Both the UFC and the NFPA codes were updated in 1994 editions and allow more tank capacity than listed in BOCA's 1993 edition. The key to the BOCA code is that it presently allows only protected tanks to be installed at nonretail facilities for private use. Industry practice shows that as long as a shop-built AST installation meets the requirements of either NFPA 30, NFPA 30A, or UFC Appendix II-F as applicable to the authority having jurisdiction, the intent of the BOCA code will be met.

The Southern Building Code Congress International Standard Fire Prevention Code

The Southern Building Code Congress International (SBCCI) *Standard Fire Prevention Code*, 1994 Edition, is used by fire officials located mainly in the southeastern quadrant of the United States. This code is updated every three years and is due to be revised in 1997. During the years between code updates, interim revisions to the code can be made if voted on by the SBCCI membership. Therefore, during certain years, the main code will differ from a more recent interim revision. The authority having jurisdiction can adopt the main code every three years, but may not recognize the iterim revisions until they are adopted into the main code during the subsequent three year cycle. This fact can lead to some confusion in the interpretation of the code, since certain jurisdictions will recognize the interim revisions while others will not. Unfortunately, the issue of ASTs is currently invovled in this confusion. The 1994 code is different from the 1995 interim revision to the code. The information that follows will highlight the 1994 Main Code and then the 1995 interim revision will be discussed.

The main points of the 1994 Standard Fire Prevention Code (SFPC) as relating to ASTs require that only protected tanks or tanks in special enclosures may be installed at private, non-retail facilities. SFPC defines a protected tank as a "thermally protected aboveground storage tank listed or approved for aboveground use" by the authority having jurisdiction. Therefore, a protected tank meeting the requirements of UL 2085 or UFC Appendix Standard A-II-F-1 shall meet this code and, alternately, a non-listed protected AST can be approved for installation by the authority having jurisdiction, even if it is not built to a nationally recognized standard.

This interpretation of the code allows the authority having jurisdiction the power to approve any protected tank application it desires, whether or not the tank system has actually been tested. Obviously, this differs from the other fire codes that have been discussed but the local authority usually will require a tank system to meet at least two codes. In any case, the best way to meet the intent of the code is to install a fire protected tank that is labeled to one of the nationally recognized standards. Table 7-5 lists other pertinent 1994 SFPC points.

Table 7-5. Standard Fire Prevention Code (SFPC) Requirements

The following requirements must be met per SFPC:
• Three feet of accessible space surrounding each tank.
• Automatic fuel shutoff devices at 90% of tank capacity.
• All openings and connections to be only through the top of the tank.
• Separation distance from buildings, property lines, and public ways not less than 5 ft.
• Fuel delivery to tanks only allowed by fixed pumps, and anti-siphon devices required.
• Capacity limits: 6,000 gallons individual or 18,000 gallons aggregate capacity.
• Fuel dispensing systems installed per NFPA 30 or 30A.
• Dispensing devices allowed to be mounted directly on top of the tank.
• No minimum separation distance of delivery vehicles from tank.
• Separation distance of 100 ft between 18,000-gallon aggregate capacity tank clusters.
• Physical barrier protection for tanks (guard posts placed within 3 ft of AST spaced 4 ft on centers and minimum 30 inches in height to withstand 12,000-pound force).
• Special enclosures for inside buildings only.
• Proof of two-hour fire rating for all tank vaults.
• Detailed plans to be submitted of the installation, describing vehicle access, secondary containment, seismic design, venting, vapor recovery, and emergency controls.

Source: Developed from SFPC.

The 1995 interim revision to the SFPC took a totally new direction in offering a level playing field for various tank technologies. In 1995, SFPC adopted NFPA 30A by reference as an interim revision to the main code. Consequently, a private, non-retail facility may now utilize unprotected ASTs with appropriate distance requirements. Refer to Chapter 6 for more exacting details on NFPA 30A. Consult your local fire officials to determine if they are recognizing the 1995 interim revision to the code. This interim revision is expected to be re-adopted as a 1996 interim revision, but expect much debate among fire officials and industry when the main code is due to be updated for the 1997 edition.

CHAPTER 8

State Regulations

The central theme surrounding all federal regulations (Clean Air Act, Clean Water Act, and others) is that the states have the responsibility to enforce the regulations, but more important, have the authority to adopt more stringent regulations than the federal regulations to combat state-specific environmental problems. In the case of the CAA, the states are required to submit to the EPA a list of areas that are considered nonattainment or do not meet national ambient air quality standards. These areas then must submit a plan to reduce the emissions from sources such as ASTs with the implementation of additional controls to bring air quality within federal standards. Therefore, it is important to examine not only federal regulations, but those of state and local governing bodies that have implemented regulations to prevent pollution of the air and water contained within their jurisdiction.

The majority of states have simply adopted the federal regulations or have made only minor changes; for example:

- Approximately 50% have adopted the CAA Subparts K, Ka, and Kb controls.
- Only 25% have adopted a spill plan different from the federal SPCC Plan.
- Some 40% have implemented, to various degrees, specific regulations on ASTs.
- About 25% have adopted regulations that vary from the federal NESHAPs program.

These statistics were taken from the *Aboveground Storage Tank Guide,*[1] which provides an informative summary of regulations pertinent to ASTs of all 50 states. The discussion of state regulations in this chapter is intended to give the reader a taste of current regulations by examining a few states that demonstrate some variance from federal standards. The discussion has been streamlined to include states that have adopted regulations of special interest to AST owners or operators, or states that accommodate a large number of ASTs which make the regulation critical to environmental protection. It is of interest to note that the states with particularly stringent regulations are typically located on major waterways, in areas that have many oil terminals which pose an environmental threat, and in areas of high population with increased air pollution.

Alaska Regulations

Alaska has extensive statute and regulatory requirements for ASTs storing petroleum and hazardous substances, which directly resulted from the Valdez oil spill of 1986. In response to the Alaska statute Oil and Hazardous Substances Pollution Control (AS 40.04), the Alaska Department of Environmental Conservation (DEC) has developed additional regulations that separate Alaska from federal and other state regulations. These statutes and regulations require that certain facilities develop contingency plans as well as adhere to stringent AST regulations.

As required by state statute AS 46.04.030, an oil discharge prevention and contingency plan must be completed by those facilities—including oil terminal, crude oil pipeline, exploration, and production facilities—where the total storage capacity is greater than 10,000 barrels of noncrude or 5,000 barrels of crude oil. The facility plan requirements given in the state statutes and regulations (18 ACC 75.400 et seq.) should be reviewed for exact details, which are very similar to the OPA of 1990. The facility plan that is developed must be approved by the DEC. As part of the prevention and contingency plan, 18 ACC 75.300 outlines requirements for discharge notification to the DEC. Notification must be given as soon as a person has knowledge of a discharge of oil into water, or 55 gallons of oil solely to land outside of an impermeable secondary containment. Within 48 hours, a facility must give notification of a release of between 10 and 55 gallons of oil solely to land or of a greater than 55-gallon release into an impermeable secondary containment. For an oil discharge of between one and ten gallons solely into land, the facility only must maintain a record of the incident; no notification is required.

The Oil Pollution Prevention Requirements (18 ACC 75.005 et seq.) were also a part of the state oil and hazardous discharge statutes (effective May of 1992), which set specific requirements for ASTs. These regulations only apply to facilities that are required to have an approved oil discharge prevention and contingency plan, as discussed above. The general regulatory requirements include maintaining records for the lifetime of the facility—specifically, the history of spills over 55 gallons, including the source, cause, and corrective action taken. The specific requirements (18 ACC 75.065), including inspection, leak detection, and cathodic and overfill protection of ASTs under this regulation, must be implemented no later than January 1, 1997. The owner or the operator of a facility must implement an inspection schedule for new and existing ASTs. The owner shall maintain and inspect oil storage according to API 653, making an inspection at least every 10 years unless API 653 requires a short or a longer inspection interval. For existing tanks, the initial inspection must occur as given in Table 8-1 and is dependent on the tank age.

A new tank shall be constructed in accordance with API 650, 12D, 12F, or 12P or another standard approved by the DEC. The AST must have cathodic protection or another system to protect the bottom from exterior corrosion when warranted, and must be equipped with a perimeter leak detection system approved by the DEC. ASTs also shall be outfitted with overfill protection such as high level alarms, or be connected to automatic shutoff pumps to alert the operator.

Table 8-1. Alaska initial inspection schedule for ASTs

Tank age[1]	Initial inspection by:
30 years and older	January 1, 1994
25 to 29 years	January 1, 1995
20 to 24 years	January 1, 1996
10 to 19 years	January 1, 1997
less than 10 years	According to API 653

Source: Alaska Department of Environmental Conservation.
[1] Age of tank determined as of May 14, 1992.

ASTs must also have secondary containment to hold the largest tank volume with additional capacity to account for precipitation. The berms or the dikes should be constructed of, or lined with, materials that are resistant to damage from the product or prevailing weather conditions, and should be designed to prevent the release of spilled oil from the containment area. New tank bottoms must be lined with an impermeable liner or a secondary tank bottom, except for tanks containing a viscous product. The DEC may require installation of monitoring wells to detect oil or hazardous substances in the groundwater.

California Regulations

California has been in the forefront of environmental regulations mainly because of severe air pollution problems and the scarcity of water resources in the state. Because of this, outside agencies often examine California regulations to get an idea of the next potential wave of AST regulations in other states. California regulations are extensive and vary between areas, but generally follow the most stringent provisions of the federal regulations. This section will examine the state's AST spill contingency plan and its clean air regulations.

The California Aboveground Petroleum Storage Act became effective January 1, 1990, and required owners and operators of petroleum ASTs to file a storage statement with accompanying fee by July 1, 1990, and to implement procedures to prevent spills, which may include groundwater monitoring. This state act required that all AST facilities submit a SPCC Plan in accordance with the federal regulations contained in 40 CFR 112 by January 1, 1991. As part of the plan, a facility must report a spill when the release is greater than one barrel (42 gallons). The Regional Water Board also has the authority to require a facility to implement monitoring of ASTs. If an AST is storing petroleum with a specific gravity of 16 degrees or lighter and has the potential to impact the groundwater, the AST facility must implement a groundwater monitoring system using either an approved means of release detection or a board-approved equal.

Ozone nonattainment areas exist throughout most of California, which requires RACT standards on ASTs to eliminate VOC emissions. The California Air Resources Board oversees 31 different district boards that write and administer RACT standards to reduce air emissions. The two major districts cover the Los Angeles and San Francisco areas and are known as the South Coast Air Quality Management District (SCAQMD) and the Bay Area Air Quality Management District (BAAQMD), respectively. For the most part, the regulations produced by these two districts are based on the Kb controls administered by the EPA and generally contain more stringent details than the Kb controls for all aspects of ASTs. Because discussion of these differences from the Kb controls would require a comprehensive review, this section focuses only on major differences.

Both district regulations require all new and existing ASTs storing organic liquids to have controls largely based on the Kb controls. The main difference applies to ASTs operating at less than the capacity and vapor pressure limits set by the Kb controls. The SCAQMD regulations decreased the lower limit vapor pressure from 4.0 psia to 1.5 psia for ASTs with a capacity between 20,000 gallons and 40,000 gallons. The upper limits and the type of controls for all other ASTs are generally equal to those of the Kb controls. The regulations also set controls for ASTs, installed after January 6, 1976, storing gasoline with a capacity of between 250 gallons and 20,000 gallons. These ASTs must either have a vapor recovery system or be equipped with a pressure/vacuum relief vent set within 10% of the maximum working pressure. The BAAQMD regulation variances from Kb controls are the same as the SCAQMD regulations for ASTs storing organic liquids greater than 20,000 gallons. ASTs with a capacity of less than 40,000 gallons, containing volatile liquids with a vapor pressure greater than 0.5 psia, must have either a submerged fill line or a vapor control device. When an AST contains a volatile liquid with a vapor pressure greater than 1.5 psia and has a capacity of 10,000 gallons to 20,000 gallons, the AST must have either a pressure/vacuum relief vent set at the minimum of within 10% of the maximum AST working pressure or 0.5 psig, or have a vapor loss control unit. The same type of control options must be placed on ASTs storing gasoline with a capacity of up to 20,000 gallons.

Florida Regulations

Florida has developed probably the most comprehensive AST regulations of any state, primarily because 90% of the state's drinking water comes from groundwater, which in many areas is less than five feet from the surface. Therefore, the need to eliminate leaks from ASTs and prevent them from entering the groundwater is of grave concern to the general public. The purpose of the Aboveground Storage Tank System regulations (Fla. Admin. Code 17-762) is to provide standards for the construction, installation, maintenance, registration, removal, and disposal of stationary AST systems, to apply to facilities that have storage capacities greater than 550 gallons. These regulations are administered by the Department of Environmental Protection (DEP).

Part of the AST program is storage tank system registration, contingency planning for facilities, and discharge reporting. A facility with an AST capacity of more than 550 gallons that is in service, out of service, or of unmaintained status must register and pay a registration fee that ranges up to $5,000 per year, depending on the capacity. The Pollutant Discharge Act, enacted in response to the OPA of 1990, requires facilities to prepare an FRP in accordance with the Phase II SPCC regulations (discussed in Chapter 3), or facilities greater than 10,000 gallons can prepare a contingency plan as outlined in the Florida Administrative Code (§16N-16.033). In the event of a discharge, these facilities must notify the DEP within one working day of the discharge discovery, for amounts greater than the following reportable quantities:

- The amounts allowed under CERCLA regulations.
- 25 gallons on pervious surfaces.
- 100 gallons on impervious surfaces other than secondary containment.
- 500 gallons inside impervious secondary containment.
- An amount causing the positive response of a release detection device.

In response to a discharge from an AST, the owner or the operator shall remove as much pollutant from the system as is necessary to prevent further release into the environment. The regulations also require the owner to demonstrate financial responsibility to pay for facility cleanup and third-party liability. For systems greater than 250,000 gallons, the liability shall be $3 million per incident and a $6 million annual aggregate.

The regulations also list performance standards for new ASTs in addition to existing shop-built and field-erected ASTs. New tanks must be constructed in accordance with national standards such as UL 142, API 650, or API 620. In addition, the installation of the AST must be in accordance with NFPA 30, and existing field-erected ASTs shall be internally inspected according to API 653 prior to December 31, 1999. Because of Florida's severely corrosive environment, mainly due to salty water and warm temperatures, the regulations also require that all ASTs have cathodic protection designed and constructed in accordance with API RP 651 or NACE standards. The AST should be equipped with overflow protection, which must be either a level indicator visible to the operator, a high-level alarm, or a high-level automatic pump shutoff. Manual gauging may be used on ASTs of less than 5,000 gallons, provided that the maximum liquid level is limited to 95% of the AST capacity. The highlight of the state regulations is that all AST systems must also be outfitted with secondary containment that:

- Includes the area under the AST as well as around the AST.
- Is designed to direct any release to a monitoring point.
- Contains at least 110% of the largest tank capacity.
- Uses a synthetic liner (if needed) with a permeability rate of 1×10^{-7} cm/sec or less and can contain the pollutants for 30 days.
- Has a leak detection system.

The secondary containment requirements also apply to existing ASTs, and must be implemented by December 31, 1999. In place of installing secondary containment under the AST, the interior bottom and 18 inches of the tank shell may be coated with an epoxy liner in accordance with API 652, but secondary containment still must be installed around the AST, and it must have a leak detection system.

Louisiana Regulations

In response to the OPA of 1990, Louisiana enacted the Oil Spill Prevention and Response Act (La. Rev. Stat. Ann. §30:19:2453 et seq.) in 1991 designed to help operators prevent and respond to oil spills, provide liability for removal costs and damages, and create the Oil Contingency Fund to provide for the levy, collection, and expenditure of funds as a result of an oil discharge. Louisiana is particularly vulnerable to a major oil disaster because of its large volume of stored oil, numerous production platforms, pipelines, and tanker traffic, including the Lousiana Offshore Oil Port which receives 15% of U.S. oil imports. This situation, coupled with limited highway access to these areas, creates great potential for a major oil spill event that would severely affect the state's fishing industry, which encompasses 26% of the nation's commercial fisheries.

The state also contains Clean Air Act nonattainment areas that have led to RACT standards for all ASTs above and beyond the CAA Kb controls discussed in Chapter 4. The counties in Louisiana that are nonattainment areas and subject to these standards include six parishes around Baton Rouge and the parish containing Lake Charles. A principal feature is the required controls on new and existing tanks storing any VOC with a capacity between 250 gallons and 40,000 gallons, which is below the lower-limit tank size for the federal NSPS Kb controls. According to these regulations, these tanks are required to have a submerged fill line, a vapor loss control system, or a pressure tank to prevent vapors from entering the atmosphere. New and existing ASTs storing VOC above 40,000 gallons followed the NSPS Kb controls with some minor changes and had to be within compliance by November 1991.

New York Regulations

The New York Department of Environmental Conservation (DEC) has issued regulations for the *Handling and Storage of Petroleum* (N.Y. Comp. Codes R. 6 Part 613), which includes bulk storage ASTs, and regulations for new petroleum storage facilities (N.Y. Comp. Codes R. 6 Part 614). These regulations are unique because of their requirements of secondary containment for all ASTs, inventory reporting, and leak detection for new ASTs.

The petroleum-handling regulations apply to all storage facilities with a combined storage capacity of 1,100 gallons and mainly require that ASTs be modified to include overflow protection and secondary containment, adhere to inventory reporting, and adopt strict inspection schedules. ASTs must also have a gauging system to prevent overfilling, by alerting the operator or by automatically operating a pump cutoff. Secondary containment, constructed in accordance with NFPA 30, should be installed around ASTs that could reasonably be expected to discharge petroleum into waters or have a capacity of 10,000 gallons or more. As part of the leak detection requirements, a facility must keep daily inventory records of the AST, which should include measurement of bottom water levels, usage, deliveries, losses, and gains. Recurring gains of water in bottom, apparent product losses, or gains exceeding 0.75% of the tank volume, or 7.5 gallons per thousand gallons delivered, must be reported to the nearest office of the DEC. To ensure that a leak does not develop, an AST of greater than 10,000 gallons capacity must undergo a complete external and internal inspection at the age of ten years and every ten years thereafter.

The state has also developed minimum standards for all new ASTs, covered in Part 614, that require the ASTs to be:

- Made of steel and constructed in accordance with a national standard such as UL 142 or API 650.
- Outfitted with a certified cathodic protection system with a 30-year lifetime.
- Constructed with a double bottom or underlain with an impervious barrier with a permeability rate of water equal to or less than 1×10^{-6} cm/sec.
- Equipped with monitoring between the bottom and the impermeable liner.

The regulations also list specific requirements for the installation of an AST (in accordance with NFPA), and for repairing and reconditioning existing ASTs including internal coating of the tank bottom for cases of severe corrosion.

Pennsylvania Regulations

The Storage Tank and Spill Prevention Act was enacted in 1989 and established requirements for developing a facility spill response in addition to establishing a certification program for installers and inspectors of both USTs and ASTs in Pennsylvania. The associated regulations, effective on September 21, 1991, are administered by the

Department of Environmental Resources (DER) and prohibit anyone not DER-certified from working on ASTs or USTs.

The program requires that personnel involved in tank-handling activities (installation, modification, inspection, removal, etc.) of ASTs be certified by the DER in addition to requiring that the company under contract to complete the work also be certified. To obtain a permit, an individual must pass an examination to be certified in one or more of the 28 different certification categories involved in tank-handling activities for both ASTs and USTs (24 installer and 4 inspector categories). An individual could have obtained a temporary certification, given adequate experience, but must have submitted a permit application and passed the permit examination by September 21, 1994. The permit is valid for only three years, at which time the individual must undergo and pass a DER-approved training program before the permit is extended.

The Storage Tank and Spill Prevention Act also required certain facilities to register an AST and a UST in addition to possibly filing a preparedness, prevention, and contingency (PPC) Plan. In general, all tanks with a capacity of greater than 250 gallons must register with the state and pay an annual fee of $50 for tanks up to 5,000 gallons, $125 for tanks up to 50,000 gallons, and $300 for ASTs greater than 50,000 gallons. (There are no maximum amounts for fees paid by a facility; so the annual cost of the DEC fees can be very large.) The DER then issues a sticker to be placed in each tank as verification that the proper fee has been paid. The distributor is not allowed to fill the tank unless the sticker is visible. Facilities that have a capacity greater than 21,000 gallons are required to submit a PPC Plan to the DER. The PPC Plan must include basic information about the facility and the response team, in addition to developing a network that includes emergency response agencies and users of the water 20 miles downstream of a facility. A facility must notify these groups of a release and each January send written notice of its inventory, with a summary of spills that occurred within the last year and measures implemented to prevent future releases.

Note

1. *Aboveground Storage Tank Guide*, Thompson Publishing Group, Washington DC, February 1994, Paragraph 405.

PART III

Major Manufacturing Standards of ASTs

CHAPTER 9

American Petroleum Institute Standards

The American Petroleum Institute (API) publishes numerous documents that cover many diverse aspects of the petroleum industry, with topics ranging from exploration for oil and the production and marketing of petroleum products to standards that govern the design and the installation of petroleum-related equipment. Standards for the design and the installation of aboveground storage tanks can be found in a number of different API publications. The primary standard for the design of ASTs is the API Standard 650, *Welded Steel Tanks for Oil Storage* (hereinafter known as API 650), which is the dominant focus of this chapter. The chapter also discusses other design standards, including the API Standard 620, *Design and Construction of Large, Welded, Low-Pressure Storage Tanks* (hereinafter known as API 620), the API Standard 653, *Tank Inspection, Repair, Alteration, and Reconstruction* (hereinafter known as API

653), and various other publications applicable to petroleum aboveground storage tanks.

The design of new field-erected tanks is primarily governed by API Standards 650 and 620 and varies according to tank design pressure or the pressure exerted on the liquid surface. Atmospheric storage tanks normally are designed per API 650, which governs tanks with a design pressure between atmospheric and 2.5 pounds per square inch gage (psig) pressure. Low-pressure tanks, up to 15 psig, are designed according to API 620. Tanks with a design pressure greater than 15 psig should be designed per ASME pressure vessel codes, which will be discussed in Chapter 10. API 653 governs the maintenance and the repair of tanks that have been constructed to either API 650 or API 620.

American Petroleum Institute Standard 650

A tank constructed per API 650 is the most common field-erected aboveground storage tank. This standard has certain restrictions that must be followed in tank design to ensure compliance:

- A uniformly supported flat bottom.
- Construction of welded carbon steel.
- Storage of product at a temperature less than 200°F.
- Design pressure and vacuum less than one ounce per square inch (1.0 osig).

The main body of API 650 code does not allow a tank to be designed for a pressure greater than 1.0 osig. However, the implementation of API 650 Appendix F will allow a tank to be designed up to a pressure of 2.5 psig. API 650 contains a total of 16 appendixes for the alteration of tank design from the main body. These appendixes, listed in Table 9-1, are not mandatory for a tank built to API 650. When specifying an aboveground storage tank, the purchaser must list not only the geometry and design parameters but also the appropriate required appendixes. The remaining discussion of API 650 focuses on tank design and construction aspects, including shell design, fixed and floating roof design, and testing requirements.

Table 9-1. Appendixes of API Standard 650

Appendix	Title
A	Optional Design Basis for Small Tanks
B	Recommendations for Design and Construction of Foundations for Aboveground Oil Storage Tanks
C	External Floating Roofs
D	Technical Inquiries
E	Seismic Design of Storage Tanks
F	Design of Tanks for Small Internal Pressures
G	Structurally Supported Aluminum Dome Roofs
H	Internal Floating Roofs
I	Undertank Leak Detection and Subgrade Protection
J	Shop-Assembled Storage Tanks
K	Sample Application of the Variable-Design-Point Method to Determine Shell-Plate Thickness
L	API Standard 650 Storage Tank Data Sheets
M	Requirements for Tanks Operating at Elevated Temperatures
N	Use of New Materials That Are Not Identified
O	Recommendations for Under-Bottom Connections
P	Allowable External Loads on Tank Shell Openings

Source: API 650.

Shell Design

A very important aspect of any tank design is selection of the proper shell material to prevent brittle fracture failure. Such failure may occur when the temperature of the shell material drops to a point where the material cracks under stress produced by the product in the tank. The crack may enlarge under stress and cause the tank to rip open. API 650 addresses material requirements in Section 2 by ensuring that the shell will have proper toughness at the specified Minimum Design Metal Temperature (MDMT). The MDMT of a given area is determined by adding 15°F to the lowest one-day mean temperature, as shown in API 650 Figure 2-2. The ambient air temperature at a given time may be less than the MDMT, but the MDMT is used because it most accurately depicts the temperature of the product, which directly governs the tank shell temperature. The temperature of the product will not adjust to hourly ambient air temperature changes, but instead will reflect the mean temperature over a 24-hour period.

API Standard 650 designates the proper material group, using both the MDMT and API 650 Figure 2-1, which shows the minimum permissible design metal temperature based on the shell plate thickness for a given material. A material group can be used when the MDMT is above the line of the minimum temperature versus thickness given in API 650 Figure 2-1. The standard does allow for the use of a material group with an MDMT below the requirements given in API 650 Figure 2-1. However, the material must demonstrate adequate toughness at the MDMT, which is achieved by conducting impact tests on three shell material samples at or below the actual MDMT.

Once the proper material group is selected, the shell plate thickness is calculated. A specific material is chosen from each group given in Table 2-3 of API 650, with the allowable stress of the material given in Table 3-2 of API 650. There are two allowable stresses listed for a given material; one value is used in designing the tank shell thickness to contain the product and the other when the tank is under a hydrostatic test loading. Based on material allowable stresses, the shell thickness is determined by one of two acceptable methods in accordance with API 650. The first is the one-foot method given in API 650 Section 3.6.3 and depicted in Table 9-2. The second is the variable-point method (API 650 Section 3.6.4), with Appendix K of API 650 providing an example of the application. The two methods are very similar and are widely used by the industry, with the variable-point method generally employed for larger-diameter tanks. For construction purposes, API 650 defines the minimum plate thickness based on the tank diameter given in Table 9-3.

Table 9-2. API Standard 650 shell plate thickness equations

Method	Shell thickness equation	Variable Definitions[1]
One-foot method	$t_d = \frac{2.6D(H-1)G}{S_d} + CA$	t_d: design shell thickness t_t: hydrostatic test shell thickness
	$t_t = \frac{2.6D(H-1)}{S_t}$	
Variable-point Upper courses[2]	$t_{dx} = \frac{2.6D(H-x/12)G}{S_d} + CA$	x: defined in API 650 Section 3.6.4.6 (usually greater than12 inches)
	$t_{tx} = \frac{2.6D(H-x/12)}{S_t}$	
Appendix A	$t = \frac{2.6D(H-1)G}{(E)(21{,}000)} + CA$	G: not less than 1.0 E: joint efficiency equal to 0.7 or 0.85.

Derived from: API 650.

[1] General variable definition—D: diameter; H: design liquid level (defined in Section 3.6.3.2 of API 650); G: specific gravity; CA: corrosion allowance; S_d: design allowable stress; S_t: hydrostatic test allowable stress.

[2] The equations included in this table are for the third course or above. API 650 Section 3.6.4 contains the equations for the first two courses.

Table 9-3. API Standard 650 minimum plate thickness

Nominal tank diameter (feet)	Nominal plate thickness (inches)
< 50	3/16
50 to < 120	1/4[1]
120 to <200	5/16
> 200	3/8

Source: API 650.

[1]6.0 millimeter plate may be substituted for 1/4" plate.

Standards other than API 650 use the material allowable stress multiplied by a joint efficiency. The joint efficiency is a measure of the weld joint's effectiveness and is often dependent upon the extent of weld seam radiographs. For both the one-foot and the variable-point methods, the joint efficiency is omitted from the equation or often is assumed to be equal to 1.0 or 100%. API 650 does not require the joints to be 100% radiographed as required by other codes. Instead, a joint efficiency of 1.0 requires only a spot radiograph.

API 650 indicates that for both the one-foot and the variable-point methods a corrosion allowance must be added to the calculated design thickness. The standard does not state what the thickness corrosion allowance should be, nor does it state exactly where the corrosion allowance should be added. API 650 simply states that consideration should be given to the effects of liquid, vapor, and atmospheric conditions on the tank with respect to the addition of corrosion allowance. If the purchaser requires corrosion allowance, the amount and the exact location should be clearly specified.

There are two main appendixes that alter either the allowable stresses or the shell design method: Appendix A, Optional Design Basis for Small Tanks, and Appendix M, Requirements for Tanks Operating at Elevated Temperatures. Appendix A defines small tanks not the basis of size but rather according to the maximum permissible plate thickness, which must be less than one-half inch. Appendix A alterations from the main standard affect MDMT requirements, allowable stress, and shell thickness calculations. The shell thickness calculation is very similar to the one-foot method, as shown in Table 9-2, used by the main body of the code, with three differences. The first is the allowable stress, which must be equal to 21,000 psi, independent of the material used, and multiplied by a joint efficiency, E, equal to 0.85 or 0.70. A joint effiency of 0.85 requires a spot radiograph, very much as in the main body of the code. A joint efficiency of 0.70 does not require any radiographs and is typically used for small tanks in which a joint efficiency of either 0.70 or 0.85 would not affect the shell plate thickness. Second, the minimum specific gravity used in the shell design is 1.0. Therefore, the thickness calculations for the hydrostatic test loading of the tank are not required. Finally, the MDMT requirements are simplified in Appendix A, such that material groups I and II are used for MDMT above –20°F and groups III and IIIA are used above –40°F. Other material groups will not provide an advantage because of the single allowable stress used in the shell design. Appendix M lists requirements for modifying the allowable stresses of the shell plate because of elevated temperatures. These requirements affect tanks designed either to the main body of the standard or to Appendix A.

After the shell thicknesses are determined, API 650 requires that the shell be checked for stability against buckling and stability against overturning due to wind loads. Tank shells that are unstable under wind loads may require stiffeners or wind girders. Section 3.9 states the requirements for determining the shell stability and requirements for sizing wind girders to stiffen the shell. If the wind speed is not specified by the customer, API 650 allows the designer to specify to the purchaser a calculated wind speed for a given shell design. In addition, if not specified by the purchaser, the calculated wind speed of a given shell design is based on the nominal plate thickness. If corrosion allowance is required for the tank, the purchaser must specify that the shell stability be checked on the basis of corroded thickness rather than nominal plate thickness. Section 3.11 of API 650 requires tanks to be checked for overturning stability. If the tank is unstable, anchor chairs may be added to prevent overturning. The stability should be checked on the basis of an empty corroded tank, including any roof dead loads acting on the shell and minus any operating pressure. It is important to note that the ninth edition of API 650 clarified that internal operating pressure should be included in the overturning stability calculations.

Besides wind-imposed loads, seismic-imposed loads are a second type of naturally imposed load. API 650 includes a recommended design procedure to resist seismic loads in Appendix E, Seismic Design of Storage Tanks. However, this appendix is not required by API 650; so the purchaser must specify that the tank is to be designed for seismic loading and reference Appendix E. The basis of seismic design for a storage tank is the calculation of a moment at the tank base that is a result of seismic forces acting on the tank and product being stored. Section E.3.1 gives the basic moment equation and the definition of variables. From the moment calculated, the tank is checked for overturning stability, and the axial stresses are examined to prevent buckling of the shell. The authors have designed tanks to other national or local codes that do not directly govern the seismic design of tanks to the extent of the equations outlined in API 650. Instead, national codes predominately discuss seismic design of buildings or other structural components. Therefore, the authors recommend that flat-bottom tanks be designed using the requirements of API 650. Care should be exercised in using values from other codes in the equations of API 650 because these values often differ between codes.

The API 650 standard has added an appendix outlining design and construction requirements for tanks constructed of stainless steel material. The main difference between carbon steel and stainless steel is the reduced allowable stress of stainless steel. The appendix addresses different allowable stresses of various stainless steel materials and the method of determining shell thicknesses based on the one-foot or Appendix A method. Because of the corrosion properties of stainless steel, the appendix also addresses handling of plates during construction, attachment of carbon steel components, and testing of the tank.

Photo 9-1 shows construction of a field-erected AST shell.

Roof Design

An open top tank requires a top angle or stiffener to maintain the tank shape because of wind loads, as discussed above. For tanks that have roofs, the top angle size is not based on wind loads but on the type of roof design being implemented. API 650 (Section 3.1.5.9) outlines the minimum top angle for a tank with a roof and classifies roofs in two categories, supported and self-supported (Section 3.10). Supported roofs are cone roof tanks in which the roof plates are supported by rafters and girders, with or without columns, as shown in Figure 9-1. The plate thickness for a supported roof is generally 0.1875 plus corrosion allowance specified by the purchaser. According to API 650, a self-supported roof is either a cone, a dome, or an umbrella roof, supported only at the periphery. These various types of roofs are shown in Figure 9-1,

PHOTO 9-1.
Construction of a field-erected AST shell. (Source: Brown-Minneapolis Tank)

along with the minimum plate thickness equations. The plate thickness ranges from 0.1875 to 0.5 inch, depending on the type of roof and the tank diameter. A cone roof designed per Section 3.10.5 is comprised of plates with a slope between 2:12 and 9:12. Dome and umbrella roofs are both dome-shaped with dome roofs being double-curvature, a true dome, whereas umbrella roofs are single-curvature, meaning rolled pie-shaped sections formed together to emulate a dome. Figure 9-1 shows the difference between the two roof types and gives the equations for the minimum plate thickness from Section 3.10.6 of API 650, which is identical for both a dome and an umbrella roof. For both of these types, the roof load includes dead and live loads and must be less then 45 psf. When the load is greater than this, the thicknesses should be increased by the square foot of the ratio of the load to 45 psf. The majority of roofs are designed with a live load of 25 psf, which is the minimum allowed by API 650. This live load accounts for any water or snow that accumulates on the roof and an adequate design for walking on the roof. The purchaser should specify what live load is required if 25 psf is insufficient.

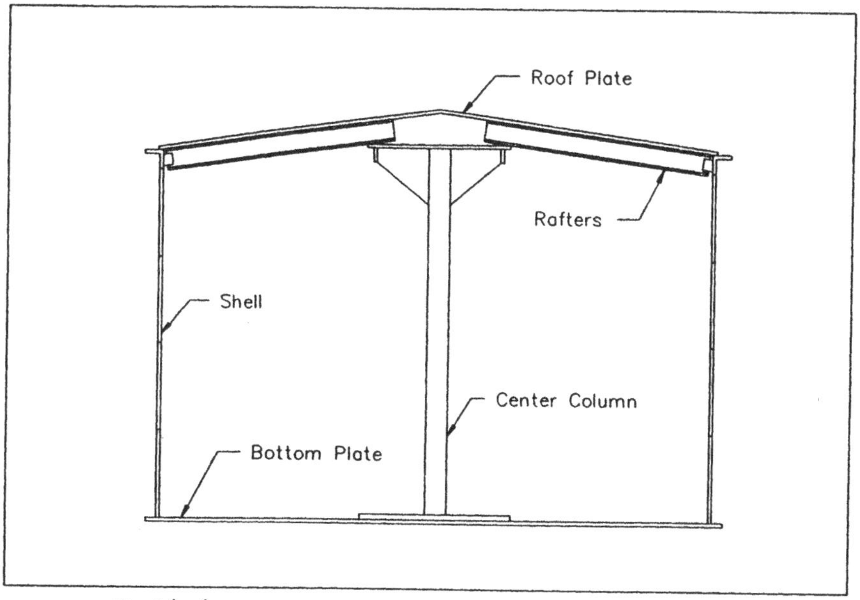

FIGURE 9-1(a).
Supported cone roof, where $\frac{3}{4}$:12 ≤ slope ≤ 2:12.

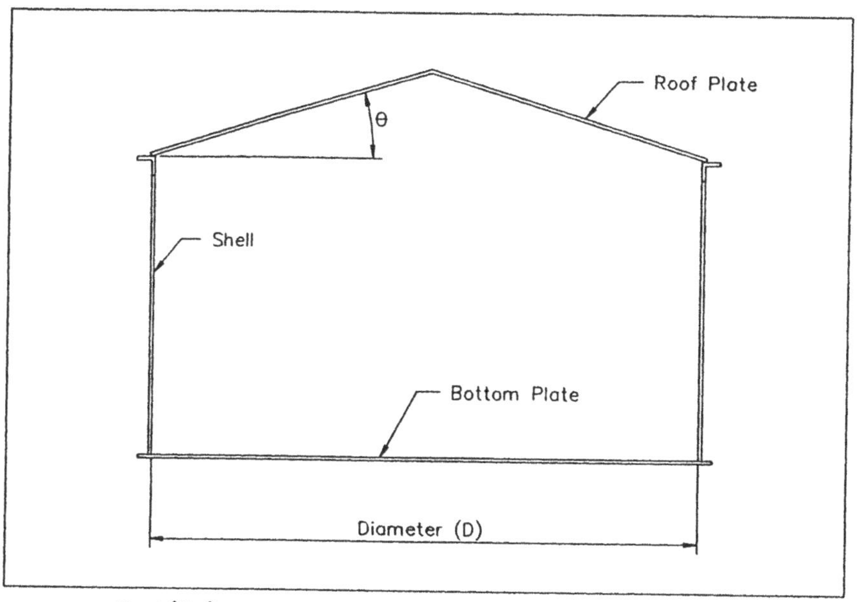

FIGURE 9-1(b).
Self-supported cone roof, where 2:12 ≤ roof slope ≤ 9:12 and $t_{min} = D \div 400 \times \sin\theta$.

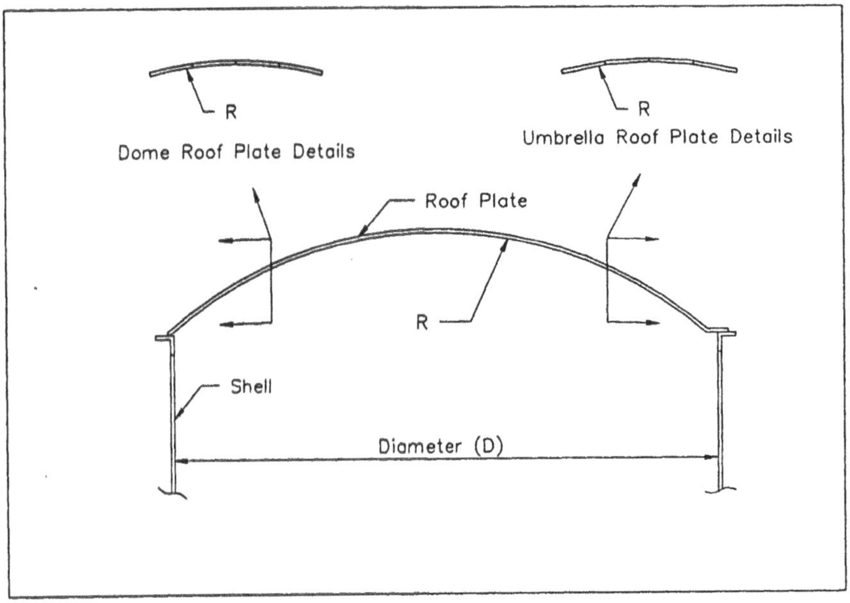

FIGURE 9-1(c).
Self-supported dome and umbrella roofs, where $0.8D \leq R \leq 1.2D$ and $t_{min} = R \div 200$.

The main body of the API 650 standard allows tanks to operate under a pressure of 1.0 osig with no additional modifications. Allowances may be made for tanks to be designed up to 2.5 psig with the implementation of Appendix F, Design of Tanks for Small Internal Pressures. The main purpose of Appendix F is to ensure that tanks have proper anchorage, and there is sufficient area at the top angle. Appendix F divides tanks into two categories, with the dividing line being where the internal pressure multiplied by the cross-section area of the tank is equal to the nominal weight of the tank. When this calculation is less then nominal weight, the tank design is as outlined in Section F.1.2. When the calculation is greater than the nominal weight, the tanks are designed per Section F.1.3.

To understand the Appendix F calculations, it is important to realize that internal pressure acts on all sides of the tank: shell, roof, and bottom. Because intersections of these elements are approximately at right angles with each other, an internal pressure will try to round out the corners and produce a smooth transition of stress between the components. When internal pressure exceeds the nominal weight of the tank, anchorage must be provided to prevent shell uplift according to Section F.7. Appendix F requires the shell and roof intersection, or top angle region, to be checked for the adequate area to provide a transition of stresses. The requirements for the top angle vary, depending upon the design section, Section F.1.2 or F.1.3, and the type of roof being used. Tanks designed per Section F.1.2, when the internal pressure does not exceed the tank weight, require that the top angle be checked per Section F.4 for any type of roof being used. For tanks outlined in Section F.1.3, the requirements are more complicated than this. For tanks with a supported cone roof, the required area is calculated according to Section F.1.2. For tanks with dome roofs or self-supported cone roofs, the roof, roof-to-shell intersection, and roof nozzles are designed per API 620 with some modifications listed in Section F.7.2 of API 650. API 650 does not allow for the design of tanks per Section F.1.3 to utilize a umbrella roof. The single-curvature roof plates produce nonparallel plate intersections and under high pressure would require reinforcement. Section F.1.3 would also require the shell to be designed for the internal pressure by increasing the design liquid level of the tank.

The pressure design equations of Appendix F are the basis for the frangible joint requirements of Section 3.10.2.5.1. When so specified by the purchaser, the tank must be checked for a frangible roof joint. The basic premise of a frangible joint is to release the internal pressure by buckling the roof-to-shell intersection rather than allowing the shell to separate from the bottom. This action would be beneficial if an explosion occurred as a result of a fire around the tank perimeter. The frangible joint requirements should be thought of more as a check on the top angle region than as a tank design parameter. Before the top angle area requirements can be checked, the roof must comply with three basic parameters:

- The weld between the roof plate and the top angle must not be greater than 3/16 inch.
- The roof plate cannot be seal-welded to the top angle on the inside.
- The roof slope can not exceed 2:12.

If these conditions are met, the top angle region must be checked to ensure that the area does not exceed the maximum allowed by

PHOTO 9-2.
Field-erected supported cone roof tanks under construction. (Source: Brown-Minneapolis Tank)

Section 3.10.2.5.3, which is dependent on the shell weight. If the area of the top angle exceeds this maximum, the tank cannot be considered frangible, and API 650 allows the tank to have emergency venting per API 2000 to relief the internal pressure.

Photo 9-2 shows field-erected supported cone roof tanks under construction.

Testing Requirements

API standard 650 outlines basic requirements for the testing of tanks to ensure a level of quality by the manufacturer. The basic testing of a tank consists of a radiograph examination of shell welds, vacuum box testing of floor seams, and testing of the tank shell. All weld joints in the shell should be full fusion and full penetration with radiograph inspection of weld seams per Section 6 of API 650, indicating the number and the location of radiographs. Section 5.3.4 outlines the various bottom weld inspections, with a vacuum box and soap suds being the most widely used. Vacuum box testing utilizes a 6 × 30-inch box with a glass top and an open-ended, gasketed bottom, which is placed over a weld joint brushed with soap. A vacuum is then applied to the box to a minimum of 2.0 psig. Any porosity in the weld joint will produce bubbles from air being pulled through. The tank shell must be tested by either a hydrostatic test, a vacuum box, or penetrating oil on the shell seams. The most common of these methods is the hydrostatic test, which consists of filling the tank with water to the maximum design liquid level. If a sufficient water source is not available, the standard allows the weld seams to be tested by the vacuum box method or by using penetration oil on one side of the shell and examining the other side for leakage.

Floating Roof Design

API 650 designates two different appendixes for the design of floating roofs: Appendix C for open top floating roofs or external floating roofs (EFR) and Appendix H for internal floating roofs (IFR). When floating roofs were first installed in storage tanks, they were used to cut vapor or product loss. Later, governmental regulations required that floating roofs be installed to curb pollution of the ambient air. An external floating roof must be designed to withstand wind and rain loading, whereas internal floating roofs are covered by a fixed roof, protecting them from the environment. Table 9-4 lists the types of floating roofs allowed for both internal and external floating roofs, with Figure 9-2 showing the various types of floating roofs.

Table 9-4. API 650 requirements for floating roofs

Type[1]	Material	IFR[2]	EFR[3]
Pan	Metallic[4]	√	√
Compartmental	Metallic[4]	√	
Annular pontoon	Metallic[4]	√	√
Double-deck	Metallic[4]	√	√
Deck-on-floats	Metallic[5]	√	
Sandwich panel	Metallic[5] or plastic[6]	√	

(continued)

Table 9-4. (continued)

Derived from: API 650.

[1] The type of floating roof is graphically described in Figure 9-1.

[2] Internal Floating Roof (IFR) is designed per API 650 Appendix H.

[3] External Floating Roof (EFR) is designed per API 650 Appendix C.

[4] Per API 650 Appendix H for internal floating roofs, material can be carbon steel, stainless steel, or aluminum; typically they are constructed from carbon steel. For external floating roofs, Appendix C only allows carbon steel.

[5] Per API 650 Appendix H for internal floating roofs, material can be carbon steel, stainless steel, or aluminum. Typically they are constructed from aluminum.

[6] Plastic sandwich-panel material is in accordance with API Standard 650, Section H.3.4 and is not typically used for internal floating roofs.

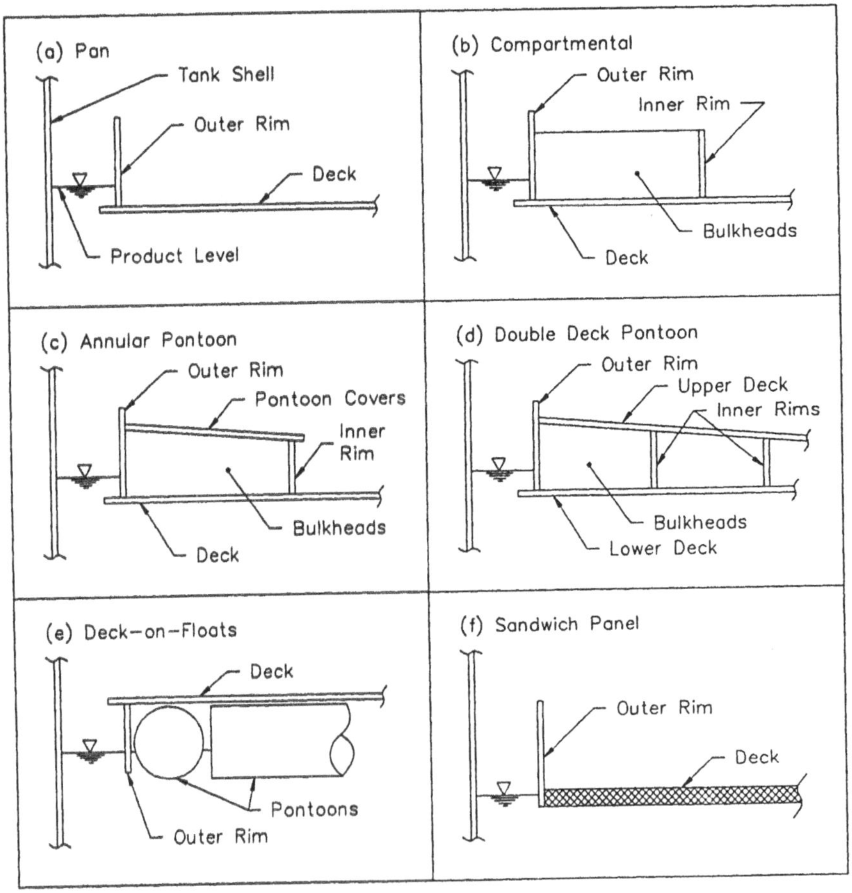

FIGURE 9-2.
Types of floating roofs.

API 650 lays out the basic design parameters for both IFR and EFR but does not describe the design procedures to follow. This is the responsibility of the manufacturer. The basic design requirement is to ensure buoyancy when supporting twice the floating roof weight. For all floating roofs except pan or sandwich-panel roofs, the buoyancy must be checked when the deck and two adjacent pontoons are punctured. The advantages and the disadvantages of the different floating roofs are discussed in Chapter 13, where the discussion shifts from the standards of the industry to the products currently available to meet those standards.

Additional Appendixes

The main body and pertinent appendixes of API 650 have been discussed, but some lesser-known appendixes need further review. The perception that API Standard 650 is a design standard just for field-erected tanks is incorrect. There exists an Appendix J, Shop-Assembled Storage Tanks, that has confused the shop-built AST market. The main difference between an Appendix J tank and other shop-built ASTs such as Underwriters Laboratories (UL) 142 tanks is that Appendix J lays out more design guidelines in line with Appendix A rather than the strict requirements for shop-built tanks per UL or STI. Tanks constructed in accordance with Appendix J have the following restrictions and minimum thickness requirements:

- Tank diameter is less than 20 feet.
- Shell plates must have full penetration butt welds.
- Bottom plates may be lap-welded with a minimum of 0.25 inch plate thickness.

The plate thickness should be determined by using the equations of Appendix A, with a minimum plate thickness of 0.1875 inch except for tanks greater than 10.5 feet, which should have a minimum plate thickness of 0.25 inch. Roofs may be self-supported cone, dome, or umbrella roofs in accordance with Section 3.10. Appendix J does allow for the elimination of the top angle and the shell-to-bottom corner welds by flanging the roof or the bottom, respectively, and butt welding to the shell. After the tank is fabricated, Appendix J allows the tank to be tested by either a hydrostatic test or an air pressure test. For the air pressure test, the tank should be pressurized to 2 to 5 psig and the weld seams brushed with soap suds to check for leaks. It is important to note that Appendix J tanks do not have any requirements for nozzles, as opposed to other API standards or UL codes. Instead, the customer must specifically indicate the nozzles required, and the manufacturer must size the tank venting.

API 650 Appendix B, Recommendations for Design and Construction of Foundations for Aboveground Oil Storage Tanks, addresses the foundation recommendations for supporting aboveground storage tanks. Because it would be impractical to cover design data for all possible situations of surface, subsurface, and climatic conditions, API 650 merely states recommended general considerations for foundation design. The main recommendation of Appendix B is that a thorough subsurface soil evaluation be performed to determine the allowable bearing pressure and expected settlements of the foundation. Based on these results and experience at the site, a foundation type should be selected to best support the tank. The options for supporting a tank per API 650 are either an earthen foundations or a ringwall foundation constructed of either crushed rock or concrete. The appendix provides guidelines for the foundation design, and Section 5.5.5 lists the required foundation tolerances. The tolerances are only for the area under the shell and are dependent upon the type of foundation being utilized. A concrete ringwall must be level within 0.125 inch in any 10 feet of circumference and within 0.25 inch around the total circumference. Any other foundation must be level within 0.125 inch in any 30 feet of circumference and 0.5 inch on the total circumference.

The environmental problem of leaking tank bottoms and the rules developed to prevent oil from seeping into the subgrade led to the addition of API 650 Appendix I, Undertank Leak Detection and Subgrade Protection, which recommends the design of tank double bottoms. A leak detection system must meet the following performance requirements as outlined in Section I.2 of API 650:

- Leaks in the bottom should be detectable by observation at the perimeter.
- The addition of electronic sensors for detection vapor or liquid is acceptable.
- Materials of construction shall be chemically resistant to the product being stored.

- The leak detection barrier permeability should not exceed 1×10^{-7} centimeters/sec.
- Any seams in the barrier should meet the same permeability requirements as the barrier.
- Material in contact with the subgrade should be protected against degradation.

The leak detection system of a tank can be broken down into two parts, the secondary bottom or release prevention barrier (RPB) and the medium in between the two bottoms or the interstitial space. The RPB is generally constructed of either a steel plate floor, a flexible membrane liner consisting of 30 mils reinforced membranes or 40 mils of unreinforced membranes, or a concrete slab. The medium in the interstitial space is generally either gravel or a steel spacer such as grating or wire mesh. Appendix I provides acceptable construction details for the implementation of a leak detection system and allows for variations if approved by both owner and manfuacturer. Figure 9-3 depicts two RPB designs. The actual combination of the secondary bottom and the medium used is based on a number of factors, including the type of foundation utilized and the cost of the different components. Discussion of RPBs and the associated leak detection system selection is presented in Chapter 13. API specifically denounces double bottoms as secondary containment of a tank. As noted by the proposed designs depicted in the appendix, the perimeter leak detection ports between the two bottoms are open to the atmosphere. Therefore, in the event of a leak in the primary bottom, the annulus space would not contain any product. Sealing of the leak detection ports would only serve to contain a leak in the primary bottom and may become part of the secondary containment system for a tank.

American Petroleum Institute Institute Standard 620

API Standard 620 is used in the design of low-pressure storage tanks having a design pressure of less than 15 psig sandwiched between atmospheric pressure storage tanks designed to API 650 and for high-pressure tanks designed according to ASME. Because API 650 and API 620 are both published by API, many similarities exist

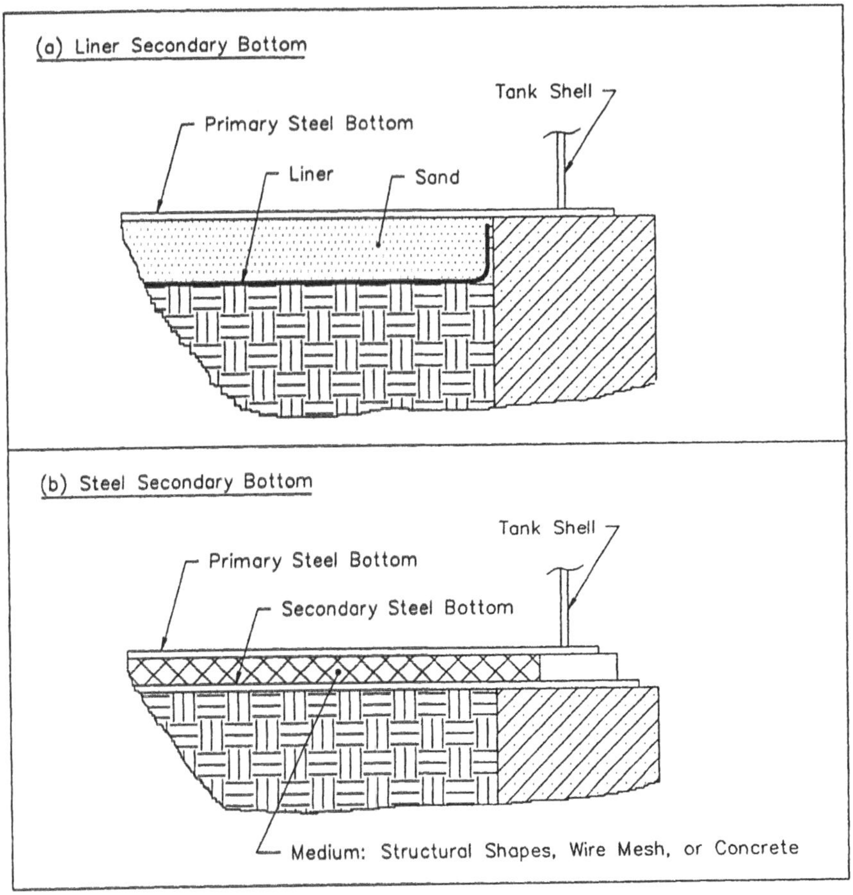

FIGURE 9-3.
Release prevention barrier designs.

between the standards. This section compares and contrasts these two standards.

The equations used to determine plate thickness per API 620 are not simplified as in API 650, but are developed from basic statics of pressure acting on a ring. This permits API 620 to be less restrictive in tank geometry, allowing the design of shells that are cylindrical, dome-shaped or double-curvature, and conical in shape. Because of this flexibility, API 620 is often used in the design of atmospheric tanks that do not fall within the guidelines of API 650. For example, an API 620 tank with a cone bottom does not require anchorage because uplift does not exist at the shell-to-bottom intersection as is found in flat-bottom tanks. The only major restrictions of API 620 tanks are that they must have a single axis of revolution, and the

PHOTO 9-3.
API 620 field-erected AST.

maximum design temepratature cannot be greater than 250°F without exception.

Photo 9-3 shows an API 620 field-erected AST.

Shell Design

Tanks must meet the toughness requirements based on the Minimum Design Metal Temperature (MDMT) specified by the purchaser. According to API 620, the MDMT shall be equal to 15°F plus the same lowest one-day mean temperature (Figure 2-1 of API 620). Exceptions to this MDMT determination are given in Section 2.2 and are allowed when the tank is insulated and heated, and when the plate stress is less than one-third the allowable tensile stress. There are no group classifications of material as found in API 650. Therefore, for any plate under pressure-imposed stresses, the material selection from Table 2-1 is based on the plate thickness and the MDMT.

The plate thickness is based on the allowable stresses given in Section 3.5. The material simple tension allowable stresses are given in API 620 (Table 3-1) with no increase allowed for hydrostatic test stresses. Because API 620 allows for the design of plates other than cylindrical shells, pressure-imposed compression stresses are common. Section 3.5.4 states that the allowable simple compressive stresses, independent of the material tensile strength, are based on the buckling capacity of the plate or the ratio of plate thickness divided by the radius. These equations are given in Table 9-5. In most cases the ratio is very small, and the simple compression allowable stress is equal to 1,800,000 times the ratio of thickness to radius.

Table 9-5. Maximum compressive stress

Thickness-to-radius ratio $(t - c)/R$	Allowable stress
$(t - c)/R < 0.00667$	$S_{cs} = 1{,}800{,}000\left[\frac{t-c}{R}\right]$
$0.00667 < (t - c)/R < 0.0175$	$S_{cs} = 10{,}150 + 277{,}400\left[\frac{t-c}{R}\right]$
$(t - c)/R > 0.0175$	$S_{cs} = 15{,}000$

Derived from: API 620.

t = wall thickness; c = corrosion allowance; R = Wall radius (inches); S_{cs} = maximum allowable simple compressive stress (psi).

Once the material has been selected and the allowable stresses are calculated, the plate thickness is determined by the equations given in Section 3.10.2. API 620 requires that the unit forces, in pounds per inch, be determined in both the meridional (T_1) and latitudinal (T_2) directions. For a cylindrical shell section, the T_1 unit force is the hoop stress created by pressure acting on a unit width of ring. The T_2 force, on a unit width of plate around the circumference, is produced by pressure acting at the ends of the cylinder. In Section 3.10.2.1, the general plate unit forces (T_1 and T_2) are given for a plate under loads created by the liquid being stored and by the gas pressure inside the tank. The simplified equations of the unit forces are given in Section 3.10.2.5 for three general cases of plate geometry: spherical or double-curvature, conical, and cylindrical. The plate thickness is determined by dividing the unit force by the allowable stress multiplied by a joint efficiency. The minimum plate thickenss is given in Table 3-6 of API 620 and is reproduced in Table 9-6. As discussed above, the allowable stresses given in Section 3.5 are "simple" allowable stresses. Both tensile and compressive stresses can exist in a plate simultaneously, a condition termed "biaxial stress." When biaxial stress is present, the allowable tensile stress is reduced

Table 9-6. API 620 minimum plate thickness

Nominal tank radius (feet)	Nominal plate thickness[1] (inches)
≤ 25	3/16
> 25 to 60	1/4
> 60 to 100	5/16
> 100	3/8

Source: API 620.

[1] Minimum plate thickness should be equal to 3/16 inch plus corrosion allowance.

according to the ratio of actual compressive to allowable compressive stress. The procedure for determining the allowable tensile stress is given in Section 3.5.3.3 of API 620, with Appendix F providing some examples of plate thickness determination

The joint efficiency needed to determine the plate thickness is given Table 3-1 of API 620, which lists the minimum joint efficiencies for various types of welds (i.e., single and double butt welds and single and double fillet welds). The joint efficiencies for double-butt welded plates can be either 85% or 100%. It is important to note the differences in joint efficiency of 100% between API 620 and API 650. Per API 650, 100% means spot radiograph, whereas in API 620, 100% means full radiograph examination of the complete length of shell seams. Because of the expense of full radiograph examination, a tank designer usually limits the joint efficiency to 85% for an API 620 tank.

API 620 requires a tank to be designed to withstand overturning due to wind or seismic loading, but does not require that the shell be checked for buckling stability under wind loads. Although Section 3.10.5 describes the limit on external pressure or vacuum-imposed loads needed to prevent shell buckling, no requirements exist for the maximum wind loads. Therefore, if the owner requires a tank to withstand a specific wind loading, this should be specified along with the method of determining shell stability. For seismic loading, Appendix L of API 620 contains the seismic loading guidelines to prevent overturning and for shell stability, which are very similar to API 650 requirements. Because of the large internal pressure of an API 620 tank, the standard addresses tank stability against overturning resulting from a combination of pressure and wind or seismic loads. Generally, an API 620 flat bottom tank will require anchorage because of the pressure. Consequently, the anchorage should be designed for the combined load case of pressure plus wind or seismic loading.

API 620 Section 3.12 outlines the design of plate intersections that are not parallel with each other and possess an internal angle of less than 180 degrees, as occurs at the intersection between the roof and the shell. The standard requires proper reinforcement at the intersection, between the roof and the shell, to provide for a transition of stresses. Under internal pressure-imposed loads, the roof plates are under tension and the intersection is in compression. Section 3.18 of API 620 outlines the requirements to be followed when the internal angle between two plates is greater than 180 degrees, as exemplified by the intersection between large cylindrical openings in cone or dome roofs. Under pressure-imposed loads on the roof, this intersection is under tension. For both of these intersection angle situations, the load produced at the intersection reinforcement is a result of the unit force (T_1) in the roof. The unit forces (T_2) in the shell and the roof resist the load produced in the reinforced intersection.

API 620 requires that the reinforcement and the attachment of each nozzle in a pressure-containing element be custom-designed as opposed to the set nozzle reinforcement sizes and specific weld details of API 650. Section 3.16 details the requirements for determining the amount of reinforcement needed and what parts can be used to reinforce the nozzle. If there exists sufficient excess area in the shell and the nozzle for reinforcement, additional reinforcement is not required, and the nozzle is considered self-reinforced. API 620 does allow a 50% reduction of the required reinforced area for external pressure-imposed loads. The nozzle weld detail must be designed with the requirements given in Figure 3-8 of API 620.

Testing Requirements

For a flat bottom tank the basic testing requirements are a vacuum-box examination of the bottom, radiograph examination of shell butt welds, and a hydrotest of the tank. The shell butt welds are examined according to the joint efficiency used to design the tank. If a joint efficiency of 100% was used, the joints should undergo a full radiograph examination per Section 5.15. If the joint efficiency is 85% then the joints should be spot radiographed per Section 5.16.

Section 5.23 outlines the leak detection testing requirements. If the tank is flat bottom, the bottom seams must be vacuum-box-tested. The tank then should be hydrostatically tested, and the vapor space above the maximum liquid level should be pressurized and checked for leaks. The pressurization of the vapor space can be accomplished by two different methods. The first is a tank hydrostatic test to the maximum design liquid level, combined with an air pressure equal to 1.25 times the design pressure and held for one hour. The pressure should be reduced to the design pressure for a close inspection of the roof seams, which is accomplished by brushing soap suds on the weld seams. The second method of testing is a tank hydrostatic test to a water level such that the pressure at the roof apex should be equal to 1.25 times the design pressure and also held for one hour. The pressure should be reduced until the pressure at the apex is equal to the design pressure, to allow a close inspection of the roof seams. During the test, extreme caution should be taken, as the pressure under the roof constitutes a large amount of energy that could be dangerous should problems arise.

American Petroleum Institute Standard 653

Three major incidents involving aboveground storage tank failures in the late 1980s led to the development of API Standard 653, *Tank Inspection, Repair, Alteration, and Reconstruction.*[1] This standard provides for repair and maintenance of field-erected aboveground storage tanks (ASTs) to address the prevention of future such accidents. An AST is defined by API 653 as a welded or riveted, nonrefrigerated, atmospheric pressure tank built to API 650 or its predecessor 12C. Although the standard is for API 650 field-erected tanks, the guidelines could be used on any AST such as an API 620 tank. It is important to note that Section 1.2 of API 653 states that compliance with this standard is the responsibility of the owner or the operator of the tank. If parts of the repair, inspection, or evaluation are contracted to other parties, the scope of the work must be well defined by the owner or the operator of the tank. Table 9-7 gives the table of contents for API 653, which can be divided into the main parts: inspection, repair, and reconstruction.

Table 9-7. API Standard 653 table of contents

Section	Title
Section 1	Introduction
Section 2	Suitability for Service
Section 3	Brittle Fracture Considerations
Section 4	Inspection
Section 5	Materials
Section 6	Design Considerations for Reconstructed Tanks
Section 7	Tank Repair and Alteration
Section 8	Dismantling and Reconstruction
Section 9	Welding
Section 10	Examination and Testing
Section 11	Marking and Recordkeeping

Derived from: API 653.

Tank Inspection and Evaluation

A tank inspection is recommended at set intervals over a tank life, when the tank shows signs of distress, or when the operator is considering a planned change of service. Section 4 states that tank inspection frequency is dependent on a number of factors, including corrosion history and prevention systems on the tank, product stored relative to the leak detection systems, and environmental impact of a discharge. An external visual inspection must be completed monthly by the operator, with a formal inspection by a qualified inspector once every fifth year or at one-quarter the corrosion life of the tank. This should include a thorough visual inspection and may include an in-service ultrasonic thickness determination of the tank shell. In addition, tank bottom and shell thickness readings are recommended to determine the tank's suitability for continued service. The interval of future internal inspections is primarily dependent upon corrosion rates, based on past inspections and on rates from tanks in similar service. This interval cannot be greater than 20 years, but shall be within 10 years if corrosion rate data are insufficient. The scope of the inspection should be determined by both a qualified inspector and the tank owner. Appendix C of API 653 has extensive checklists for both in-service and out-of-service inspections.

The primary purpose of an internal inspection is to ensure that the tank bottom is not severly corroded, leaking, or experiencing excessive differential settlement. Section 2.2.4 lists 12 historic failures of bottoms that led to leakage that may be categorized as corrosion of the top and the bottom side, differential settlement of the bottom, or improper design and construction of both the tank bottom and the foundation. Therefore, API 653 describes specific requirements for inspection and determination of continued service based on the plate thickness and defects. According to API 653 Table 4-1, the minimum bottom plate thickness should be greater than 0.10 inch for continued service if the tank has no means of leak detection or containment. If leak detection and containment do exist, or if the tank floor is lined in accordance with API 652, *Lining of Aboveground Petroleum Storage Tanks*, the minimum thickness for continued service is 0.05 inch. According to API 653 Section 2.4.7, the plate thickness evaluation is based on either of two methods: deterministic and probabilistic. The deterministic method is based on two equations containing many variables in which the minimum actual plate thickness is derived from a thorough scan of the bottom plates. The probabilistic approach is accomplished by taking random plate thickness readings and statistically determining the minimum actual plate thickness based on the sample of readings. If the bottom plate thickness falls below the minimum allowed, the bottom must be lined, repaired, or replaced. In addition to corrosion, bottom settlement can lead to high stress concentrations and cracks, which may result in further leakage. Appendix B of API 653 has requirements for measuring settlement and acceptable limits. Chapter 14 discusses methods of repair for the diferent types of settlement.

API 653 Section 2.3 describes the tank shell evaluation necessary for continued service. Besides visual inspection of the tank shell and welds for distortion and flaws, the shell inspection includes plate thickness readings completed by ultrasonic examination. Once the plate thickness is determined for a given region, the minimum plate thickness for a welded tank shell is calculated using the one-foot method. The allowable stress is slightly larger than that allowed by API 650 for a given material because the equation is dealing with small areas of reduced thickness due to corrosion and not the average plate thickness. In cases when the shell plate material or design information is unknown, API 653 gives recommended allowable stress and a joint efficiency for determination of the minimum plate thickness. If the thickness of the plate is inadequate, the tank shell should be repaired or replaced, or the maximum product level should be reduced, on the basis of various minimum plate thicknesses.

To justify continued service or when a change in tank service is planned, the risk of brittle failure must be evaluated. API 653 has devoted all of Section 3 to considerations of brittle failure assessment. The recommendations are laid out in a decision tree, Figure 3-1 of API 653, reproduced here as Figure 9-4. The risk of brittle fracture failure is minimal when the tank material type of service exhibits adequate toughness. A tank that meets at least the requirements of the seventh edition of API 650 automatically meets the toughness requirements of API 653, and the risk of brittle fracture failure is minimal. A tank also meets the toughness requirements of API 653 when a change in service does not violate API 650, or when any tank repairs are completed per API 653. If the tank does not met the requirements of API 650, as in the case of an unknown shell material, it can undergo a hydrostatic test to demonstrate toughness requirements for continued service. The tank can forgo a hydrostatic test provided that the criteria given in Figure 9-4, footnotes 4 through 8, are met.

Tank Alterations and Modifications

Another function of API 653 is to ensure that tank alterations or modifications are completed in such a way so as not to decrease the integrity of the tank. The rules for the repair and alterations of tanks are given in Section 7, which states that the basis of alteration is equivalent to the rules of API 650. Modifications to shell nozzles and shell plate replacement are tank alterations that must be completed with adequate clearance between existing nozzles and seams. Sections 7.6 and 7.7 list additional requirements for nozzle modifications that follow the guidelines of API 650. Figure 7-1 of API 653, reproduced here as Figure 9-5, is a graphical definition of the weld clearance requirements of shell plate replacement. The minimum dimension of a shell replacement plate must be 12 inches or 12 times the shell plate thickness. Replacement plates must intersect the horizontal seams at approximately 90 degrees or have rounded corners. Full-height shell plate replacement is allowed, but a minimum distance from existing weld seams must be maintained.

Another modification commonly completed on a tank is an entire bottom replacement, which must be conducted according to Section 7.9 of API 653. These requirements call for the shell to be slotted and new plates extended out beyond the shell in compliance with API 650. The annulus space between the bottoms should be three to six inches thick and contain a suitable noncorrosive material such as sand, gravel, or concrete. When the new bottom is installed three to six inches above the existing floor, the existing shell penetrations may need to be raised. The work should be completed per Section 7.8 of API 653 to provide the required weld spacing of API 650.

Testing Requirements

To ensure proper quality in any repair or alteration, Section 10 of API 653 lists specific testing requirements similar to those of API 650. During repair or alteration of a tank, the length of down time is crucial to the operations of a facility. The tank hydrostatic test can take several days and thus is a major factor in the down time duration. Section 10.3 discusses the hydrostatic test requirements for a tank and states that a reconstructed tank or a tank with major repairs must be subjected to a 24-hour hydrotest. The standard defines a major repair as one in which a large portion of the shell below the design liquid level is altered, the annular ring is modified, or the corner weld is repaired. API 653 lists a number of major repairs, which are summarized below:

- Installation of a 12-inch or larger shell penetration.
- Replacement of shell plate or annular ring plate.
- Repair of more than 12 inches of vertical shell seam or radial annular ring joints.
- Installation of a new bottom.
- Replacement of the corner weld.
- Jacking of the tank shell.

Omission of the hydrostatic test requirements for tanks that have undergone major repairs is permitted if the conditions given in Table 10-1 of API 653 are met. These exceptions exist if the tank has a relatively thin shell based on the design metal temperature given by API 653 Figure 3-2, the tank shell meets the toughness requirements of API Standard 650, seventh edition or later, or the repairs are limited to the tank bottom, away from the corner weld. However, these exceptions are currently being reviewed and will likely change in 1995. For these conditions of hydrostatic test omission, API 653 states that the repair work done on the tank must be approved by an experienced storage tank engineer, and all work must be in accordance with the requirements of API 650. The standard also requires that any repairs made to the shell or shell penetrations have additional testing as conveyed in Table 10-1 of API 653.

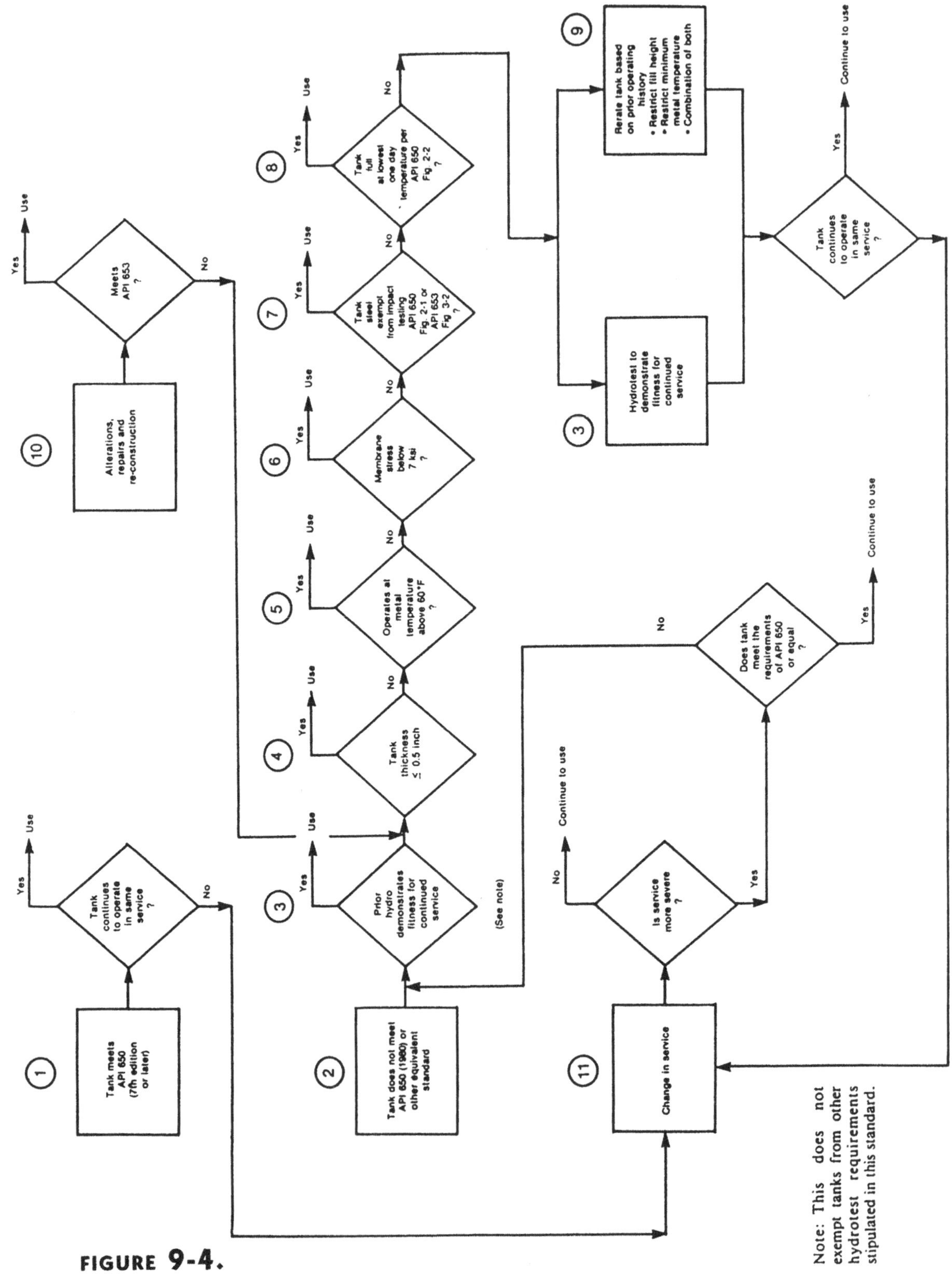

FIGURE 9-4.
Brittle fracture considerations. (Source: American Petroleum Institute)

The assessment procedure as illustrated in Figure 9-4 shall be used for evaluating aboveground atmospheric storage tanks in petroleum and chemical services. Each of the key steps on the decision tree are numbered corresponding to the explanation provided as follows:

1. These tanks meet the API Standard 650, seventh edition or later, requirements to minimize the risk of failure due to brittle fracture. Tanks may also be shown to meet the toughness requirements of API Standard 650, seventh edition or later, by impact testing coupon samples from a representative number of shell plates.
2. Many tanks that continue to operate successfully in the same service were not built to the requirements of API Standard 650, seventh edition or later. These tanks are potentially susceptible to failure due to brittle fracture and require an assessment as illustrated by the decision tree.
3. For the purpose of this assessment, hydrostatic testing demonstrates that an aboveground atmospheric storage tank in a petroleum or chemical service is fit for continued service and at minimal risk of failure due to brittle fracture, provided that all governing requirements for repairs, alterations, reconstruction, or change in service are in accordance with this standard (including a need for hydrostatic testing after major repairs, modifications or reconstruction). The effectiveness of the hydrostatic test in demonstrating fitness for continued service is shown by industry experience.
4. If a tank shell thickness is no greater than 0.5 inch, the risk of failure due to brittle fracture is minimal, provided that an evaluation for suitability of service per Section 2 has been performed. The original nominal thickness for the thickest tank shell plate shall be used for this assessment.
5. No known tank failures due to brittle fracture have occurred at shell metal temperatures of 60°F or above. Similar assurance against brittle fracture can be gained by increasing the metal temperature by heating the tank contents.
6. Industry experience and laboratory tests have shown that a membrane stress in tank shell plates of at least 7 ksi is required to cause failure due to brittle fracture.
7. Tanks constructed from steels listed in Figure 2-1 of API Standard 650 can be used in accordance with their exemption curves, provided that an evaluation for suitability of service per Section 2 of this standard has been performed. Tanks fabricated from steels of unknown toughness thicker than ½ inch and operating at a shell metal temperature below 60°F can be used if the tank meets the requirements of Figure 3-2. The original nominal thickness for the thickest tank shell plate shall be used for the assessment. For unheated tanks, the shell metal temperature shall be the design metal temperature as defined in 2.2.9.3 of API Standard 650.
8. The risk of failure due to brittle fracture is minimal once a tank has demonstrated that it can operate at a specified maximum liquid level at the lowest expected temperature without failing. For the purpose of this assessment, the lowest expected temperature is defined as the lowest one day mean temperature as shown in Figure 2-2 of API Standard 650 for the continental United States. It is necessary to check tank log records and meteorological records to ensure that the tank had operated at the specified maximum liquid level when the one-day mean temperature was as low as shown in Figure 2-2 of API Standard 650.
9. An evaluation can be performed to establish a safe operating envelope for a tank based on the past operating history. This evaluation shall be based on the most severe combination of temperature and liquid level experienced by the tank during its life. The evaluation may show that the tank needs to be rerated or operated differently; several options exist:
   a. Restrict the liquid level.
   b. Restrict the minimum metal temperature.
   c. Change the service to a stored product with a lower specific gravity.
   d. Combinations of a., b., and c., above.

   The owner/operator can also make a more rigorous analysis to determine the risk of failure due to brittle fracture by performing a fracture mechanics analysis based upon established principles and practices. The procedures and acceptance criteria for conducting an alternative analysis are not included in this standard.
10. All repairs, alterations and relocations shall be made in compliance with this standard.
11. An assessment shall be made to determine if the change in service places the tank at greater risk of failure due to brittle fracture. The service can be considered more severe and creating a greater risk of brittle fracture if the service temperature is reduced (for example, changing from heated oil service to ambient temperature product), or the product is changed to one with a greater specific gravity and thus increasing stresses.

FIGURE 9-4.
footnotes

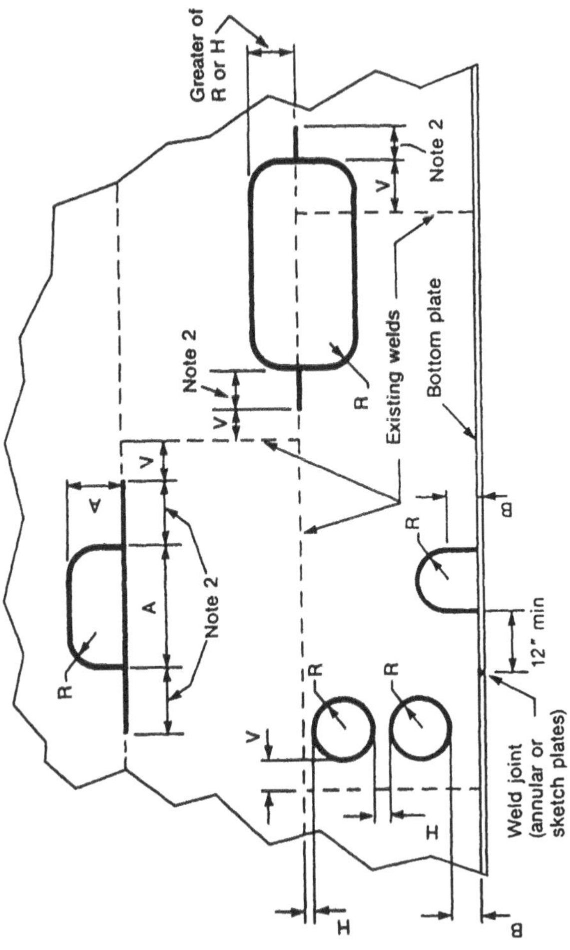

Dimension	Minimum weld spacing between edges (toes) of welds for thickness of replacement shell plate, t, (inches)	
	$t \leq 0.5$ inch	$t > 0.5$ inch
R	6"	Greater of 6" or 6t
B	6"	Greater of 10" or 8t
H	3"	Greater of 10" or 8t
V	6"	Greater of 10" or 8t
A	12"	Greater of 12" or 12t

Notes:
1. All weld intersections shall be at approximately 90°.
2. Prior to welding new vertical joints, cut existing horizontal weld for a minimum of 12" beyond the new vertical joints. Weld the horizontal joint last.

FIGURE 9-5.
Acceptable details for replacement of shell plate materials. (Source: American Petroleum Institute)

American Petroleum Institute Recommended Practice 651

The API Recommended Practice 651, *Cathodic Protection of Aboveground Storage Tanks* (hereinafter known as API 651), presents recommend practices for the design of cathodic protection systems for aboveground storage tanks. The focus of API 651 is to outline the information needed to design a system and to describe the different systems currently available for installation on tanks. Section 2 describes the corrosion of aboveground storage tanks, including that due to stray currents, bimetallic corrosion, and internal corrosion. Stray currents in the soil travel through the tank and cause corrosion due to loss of electrons at the point where the current reenters the soil. Examples of stray current sources include DC electric power sources such as power lines, electric trains, and welding machines and impressed current cathodic protection systems for pipelines. Bimetallic corrosion results from the loss of electrons between two metals of different electrolyte potentials. An example would be the installation of copper or stainless steel piping near a steel tank. The potential difference would cause current to flow from the tank to the dissimilar metal, resulting in electron loss and corrosion. Finally, internal corrosion results from the storage of corrosive products such as acids, or in tanks storing petroleum products which may corrode the shell at a water–petroleum interface.

A number of factors influence the need for corrosion protection as outlined in Section 3 of API 651, including soil characteristics, construction and cathodic protection of the tank, tank history, and location of other tanks in the tank farm. Critical to the corrosion protection of a tank are the foundation design and construction. The AST should have adequate drainage to prevent collection of water under the tank. In addition, the resistivity of the soil supporting the bottom should be made as high as possible, as greater soil resistivity will lead to a smaller current flow of electrons from the tank bottom. Soil imperfections should also be considered because debris, such as clay, rocks, and weld rods which are excellent conductors of electricity, may lead to accelerated corrosion of the bottom plate. According to API 651, the use of a properly drained concrete or asphalt base isolates the tank from the soil and provides adequate cathodic protection for the bottom.

When an evaluation of the tank and the soil warrants cathodic protection, API 651 recommends either a sacrificial anode or an impressed current system. The sacrificial anode is a bimetallic corrosion system where the sacrificial anode is shifted from the tank to a more active metal such as magnesium or zinc. The anode gives up electrons to the tank, causing corrosion at the anode rather than the tank. These anodes are buried under the tank bottom, where sand completes the circuit, allowing electron movement to the tank. Because a sacrifical system has limited current output, the system is most economical for small tanks or tanks protected with a lining. An impressed current system would be most economical on a large tank or one with a bottom that is bare or poorly coated. An impressed current system works much the same way as a sacrificial anode system except that there is a rectifier attached to the sacrifical anode that drives the protection system. The rectifier reduces the AC current coming into the system and converts the power to a DC source. Table 9-8, taken from API 651, lists the advantages and disadvantages of both systems. The design of the cathodic system is discussed in Section 6 of API 651, and it should be completed by an engineer familiar with the design of cathodic protection systems.

American Petroleum Institute Recommended Practice 652

Another publication issued by the API to aid environmental protection is API Recommended Practice 652, *Lining of Aboveground Petroleum Storage Tanks* (hereinafter known as API 652). The intended purpose is to provide procedures for the installation of linings to protect the tank bottom from internal corrosion. According to API 652, petroleum products have a corrosion rate of two mils per year, which is very low. Corrosion rates can accelerate with the settlement of water or other chemicals to the bottom of a petroleum storage tank. The need for tank bottom corrosion protection may not be known until after the tank has been in service for some time. Therefore, the need for protection would be based on tanks of similar design, storing similar products, or based on governmental regulations. According to API 652, there are two types of linings utilized on tank bottoms: thin-film and thick-film. Thin-film linings are 20 mils or less and comprised of epoxy or epoxy-copolymer resins. The application of thin films is usually on new tanks where the bottom is

Table 9-8. Advantages and disadvantages of sacrificial anode and impressed current systems

System	Advantages	Disadvantages
Sacrificial anode	• No external power supply is required. • Installation is relatively easy. • Capital investment is low for small-diameter tanks. • Maintenance costs are minimal. • Interference problems (stray currents) are rare. • Less frequency monitoring is required.	• Driving potential is limited. • Current output is low. • Method is limited to use in low-resistivity soils. • Method is not practical for protection of large structures.
Impressed current	• Availability of large driving potential • High current output capable of protecting large structures. • Capability of variable current output. • Applicability to almost any soil resistivity.	• Interference problems (stray currents) on foreign structures. • Loss of AC power cause of loss of protection. • Higher maintenance and operating costs. • Higher capital cost for small installations. • Safety aspects of rectifier location. • Safety aspects of negative lead connection. • More frequent monitoring

Source: API 652.

smooth. Thick-film linings are glass-reinforcement systems with a few generic linings listed in Table 2 of API 652. Thick-film linings have sufficient strength to bridge over small imperfections, but are brittle so that cracking of the lining can be a problem. In addition, the costs of thick-film linings are typically high because of their difficult installation.

Additional American Petroleum Institute Tank Standards

Vapor Losses

The American Petroleum Institute issues several publications that calculate estimated vapor losses from storage tanks (see Table 9-9). When they were first issued, their purpose was to estimate product loss as it correlates with lost profits. Their use has shifted as environmental concerns about vapor loss have become of increasing public importance. Therefore, these publications are used in obtaining construction permits that require estimates of total vapor losses from tanks and their effects on ambient air quality. These publications should be used to determine the type of controls required on tanks to ensure that the tank emissions do not exceed those allowed by the air pollution requirements, which are discussed further in Chapter 13.

Table 9-9. API storage tank evaporation loss publications

Publication	Title
API Bulletin 2513	Evaporation Loss in the Petroleum Industry—Causes and Control
API Bulletin 2516	Evaporation Loss from Low-Pressure Tanks
API Bulletin 2518	Evaporation Loss from Fixed-Roof Tanks
API Bulletin 2519	Evaporation Loss from Internal Floating-Roof Tanks
API Publication 2517	Evaporation Loss from External Floating-Roof Tanks

Source: API.

Additional Tank Standards

There is some confusion in the industry as to the role different standards and codes play in specification of a tank. Generally, API is considered to produce standards for field-erected tanks, and UL and STI provide standards for shop-built tanks. As discussed above, API publishes as part of API 650 an appendix for the design of shop-built tanks, Appendix J. This appendix is not a code such as UL 142, which lists specific requirements for tank size and appurtenances, but rather a guideline used to design tanks. API publishes two additional codes that come close to the code layout of UL 142: API Specification 12F, *Specification for Shop Welded Tanks for Storage of Production Liquids*, and API Specification 12D, *Specification for Field Welded Tanks for Storage of Production Liquids*. As stated in the forewords to both standards, the purpose is to offer convenience to both the purchaser and the manufacturer when ordering a tank. These specifications were originally designed to allow select manufacturers to build these tanks and affix an API monogram on the nameplate. Currently, any manufacturer can build API 12F and 12D tanks.

The API 12F standard specifies the tank dimensions and required nozzle size and location in Table 3.1 and Figure 3.1 of the standard. The size of tanks listed in the table range from 90 barrels (7′11″ in diameter by 10′0″ high) to 750 barrels (15′6″ in diameter by 24′0″ high). The standard also lists specific construction details of the tank. The bottom shall be flat or coned to the center with the foundation contoured to match the tank bottom, or the tank shall be constructed on a short skirt, 14 inches high. The bottom plates are double-butt welded with the bottom at the shell intersection either flanged or a double-fillet welded T-joint. These shell-to-bottom intersections match the types allowed by the Appendix J requirements of API 650. The shell also should be double-butt welded and constructed of 0.1875-inch plate or 0.25-inch plate as specified by the purchaser. In addition, the roof should be double-butt welded with a slope of one in twelve. The tank does not require a top angle with the roof-to-shell intersection identical to the shell-to-bottom intersection with a flanged roof or a T-joint. To check for leaks in an API 12F tank, the standard requires that the tank be pressurized to 1.5 or 3.0 psig, depending on the tank size, and that weld seams be brushed with soap suds.

Like API 12F, API 12D for field-erected tanks also has specific sizes and nozzles, which are listed in Table 3.1 and Figure 3.1 of the standard. The sizes of tanks range from 500 barrels (15′6″ in diameter by 16′0″ or 21′6″ in diameter by 8′0″ high) to 10,000 barrels (55′0″ in diameter by 24′0″ high). Except for the specific nozzle size and tank dimensions, the requirements of this standard are very similar to those of API 650 Appendix A. The bottom is constructed of 0.25-inch plate, which can be single lap welded with the shell-to-bottom connection a double-fillet welded T-joint, but the bottom is not allowed to be flanged. The tank roof may be flanged, and the standard requires a top angle. For roofs greater than 15′6″ in diameter, internal structural rafters are required to support the roof loads. After construction, the tank should be tested first by vacuum box testing the bottom seams and then by filling the tank with water.

American Petroleum Institute Standard 2610

The API has taken their experience in the design and the operation of ASTs and published a standard that incorporates all aspects of the AST, from initial planning to final operations. Recently, a committee called Terminal and Tank Standards Working Group (TTSWG) set out to examine and recommend changes to the current API publications. The purpose was to take a proactive role in examining the issues that face AST management and recommend changes to publications in an attempt to eliminate the need for unnecessary legislation. The committee has taken into account the full spectrum of the petroleum industry, including petroleum terminals, refinery facilities, bulk plants, asphalt plants, and aviation service facilities. To meet the needs of these AST owners and operators, the TTSWG produced API Standard 2610, *Design, Construction, Operation, Maintenance, and Inspection of Terminal & Tanks Facilities* (hereinafter known as API 2610) to guide the management of terminals in implementing safeguards for the public and the environment.

Table 9-10. API Standard 2610 table of contents

Section	Title
Section 1	General
Section 2	Site Selection and Spacing Requirements
Section 3	Pollution Prevention and Waste Management
Section 4	Safe Operations of Terminals & Tanks
Section 5	Fire Prevention and Protection
Section 6	Tanks
Section 7	Dikes and Berms
Section 8	Pipe, Valves, Pumps, & Piping Systems
Section 9	Loading, Unloading, and Product Transfer Facilities
Section 10	Corrosion Control
Section 11	Structures, Utilities, and Yard
Section 12	Removals and Decommissioning of Facility

Derived from: API 2610.

Discussion will focus only on the API 2610 highlights; Table 9-10 is the table of contents. Sections 2 and 3 of API 2610 focus on the location of a facility so as to protect the public safety and the environment. The standard references the appropriate fire codes to properly space ASTs from existing structures and the appropriate federal regulation requirements to meet the pollution prevention needs of a facility.

Sections 4 and 5 of API 2610 address the safe operations of an AST, which includes OSHA regulations and NFPA regulations for proper fire protection. The standards reiterate the need to identify the product being stored using MSDS sheets and provide safe working practices around ASTs to include proper training and facility inspection. Section 5 mainly references the appropriate section of the NFPA and the other pertinent fire codes for the design and operations of a fire protection system.

Section 6 of API 2610 focuses on the specific requirements of the AST, recommending that the ASTs be constructed in accordance with the applicable API standards discussed previously in this chapter. This section assists the facility management on AST selection, based on its planned service, and discusses proper environmental features to include floating roofs, release prevention barriers, or leak detection. In addition, the standard recommends the type of apertures that should be considered for ASTs so as to to ensure proper venting, discusses the type of level gauges that can be installed, and outlines proper design for tank alarms and other pertinent items for ASTs. This section also recommends that all ASTs be inspected in accordance with API 653 and outlines nineteen points of inspections or checks.

Section 10 focuses on the corrosion protection requirements of not only the ASTs but also the associated piping. The corrosion protection discussed in this section pertains to the proper coating system, both internal and external, in addition to the design of the cathodic protection system. For coating systems, the discussion is on system selection, application procedures, and inspection and maintenance of the coatings. For the cathodic protection system, the standard references both API 651 and the appropriate NACE standards for the design and installation.

Note

1. *The Evolution and Use of API Standard 653*, by Will Carter, Aboveground Tank Management 1991–92 Reference Guide, CEEM Publications, Fairfax, VA, 1991, pp. 129–137.

CHAPTER 10

National Design Standards of Field-Erected Tanks

In addition to those of the American Petroleum Institute, a number of other publications are used in the design and the construction of aboveground storage tanks. Publications of the American Water Works Association (AWWA) and the American Society of Mechanical Engineers (ASME) are widely used, for example. AWWA publications focus on the design of tanks utilized for the storage of water. Although the purpose of most AWWA tanks is to serve local municipalities, they also are used in the design of waste treatment facilities, both municipal and industrial. ASME publishes a code for the design of pressure vessels or tanks under high pressures, above 15 psig, which are outside the limits of the API publications. For the petrochemical industry, pressure vessels are primarily used in the refining of petrochemical products. This chapter discusses these two different associations and their relevant publications for design and construction of field-erected storage tanks.

American Water Works Association

The American Water Works Association (AWWA) in conjunction with the American Welding Society (AWS) publishes a standard for the design of water tanks: D-100 (AWS D5.2), *AWWA Standard for Welded Steel Tanks for Water Storage*. The standard is commonly referred to as D100-84, in reference to the most recent standard, published in 1984. The AWWA has published a number of standards that supplement this design code, one of which is AWWA D102, previously titled *Standard for Painting Steel Water-Storage Tanks*, which is no longer in print and will be rewritten. Another standard is AWWA D105, *Standard for Disinfection of Water Storage Facilities*.

AWWA D100 Standard

The AWWA D100-84 standard is used almost exclusively in the design and the construction of water tanks and governs the design of water reservoirs, standpipes, and elevated storage tanks. The AWWA defines a reservoir as a tank where the height is less than or equal to the diameter and a standpipe as a tank in which the height is greater than the diameter, and an elevated tank is a tank supported off the ground. Although all three are used for water storage, the primary function of a standpipe is to maintain a set pressure in a water line system, and it can be very small in diameter. This chapter focuses on the design and the construction of flat-bottom tanks, which are equivalent to AWWA reservoirs and standpipes.

Since 1973, AWWA D100-84 has governed the design of welded shell plates only after the elimination of riveted shell joints. The shell plate thickness calculation and the material selection are very similar to the requirements of API 650 Appendix A. Per AWWA D100-84, the shell design utilizes a single stress value with no material toughness requirements for the material. The equation of the shell plate thickness is given below:

$$t = \frac{2.6h_pDG}{sE} + CA \qquad (10\text{-}1)$$

where:

h_p = height of liquid from overflow elevation to bottom of shell courses under design

D = tank diameter
G = specific gravity, equal to 1.0 for water
s = allowable stress
E = joint efficiency, equal to 0.85 for butt welded plates under tension per Table 9 of AWWA D100-84
CA= corrosion allowance specified by the purchaser.

According to Table 1 of AWWA D100-84, the allowable stress for shell plate under tension is equal to 15,000 psi. Because of this low allowable stress, the standard does not have any material toughness requirements, and any grade material referenced in Section 2.2.3 can be used. The standard requires that the purchaser specify the corrosion allowance and the tank components to which the corrosion allowance applies. Section 3.9 requires that the shell corrosion allowance be added to the calculated shell thickness, as shown in the equation, and to the minimum plate thickness given in Section 3.10. The minimum thicknesses are reproduced in Table 10-1 and are identical to the API Standard 650 minimum discussed in the previous chapter. The difference from API 650 is the addition of corrosion allowance to the minimum shell plate thickness.

Table 10-1. AWWA D100-84, minimum plate thickness

Nominal tank diameter (feet)	Minimum plate thickness[1] (inches)
< 50	3/16
50 to < 120	1/4
120 to < 200	5/16
> 200	3/8

Source: Derived from AWWA D100-84.

[1]Any specified corrosion allowance should be added to the minimum thickness.

Once the shell plate thicknesses have been determined based on the product stresses, the shell buckling stability against wind loading should be checked. Section 3.5 of AWWA D100-84 states rules governing the design of top wind girders on open-top tanks and gives requirements for the location of wind girders on tanks subject to wind buckling. According to AWWA, the wind speed should be taken as 100 mph unless a higher wind speed is specified by the purchaser. The standard does not allow a tank shell to be designed for a wind speed less than 100 mph as API 650 does. For flat-bottom tanks,

AWWA does not specifically require that the tanks be checked for overturning as a result of the wind loads. However, based on good engineering judgment, the authors recommend that tank overturning stability be verified in a manner similar to the method given in Section 3.10 of API 650.

The majority of tanks built according to AWWA are used for the storage of potable water. This requires that most tanks be constructed with some type of roof, usually a supported cone or an umbrella roof. As AWWA D100-84 lists only the roof loading and allowable stresses of the elements, the design procedures of API 650 generally are used for the design of tank roofs where the live load is equal to the snow load. AWWA recommends that the snow load be 25 pounds per square foot (psf) for roofs with less than a 30-degree slope. With a slope steeper than 30 degrees, the roofs are not required to be designed for any snow loads. For tanks located in areas where the lowest one-day mean temperature is 5°F or higher, the snow load does not need to be greater than 15 psf.

The AWWA standards do not require that the roofs be painted except in areas under the overflow elevation. If the purchaser requires painting of the roof underside for corrosion protection, there are some important factors to consider in specifying the roof type. First, Section 8.7 of AWWA D100-84 requires that lap-welded roofs, not under hydrostatic pressure, be welded from the top side only, a practice that leaves unprotected areas between plate laps. The purchaser can specify that these seams be seal-welded, but for supported cone roofs this is difficult and costly. Instead, the purchaser may want to consider eliminating the roof structural members and specifying an umbrella roof. Second, seal welding of rafters to the roof plates is not recommended to facilitate application of a paint system. The roof plates will fluctuate as a result of wind blowing across the roof or temperature variations in the roof plate. If the rafters are welded to the roof plates, movement of the roof plates may pick-up the columns and shift the column bases off center, resulting in column buckling due to eccentric loading. For these reasons, careful consideration should be made in the specification of the roof and painting system to ensure that the internal paint will perform as expected.

The AWWA approach to seismic design uses two methods for determining the seismic force acting on the tank. According to AWWA D100-84, the purchaser must specify the condition of seismic loading as 1, 2, or 3, which translates into no seismic loading on the tank, moment based on the fixed percentage method, and moment based on the pseudodynamic method, respectively. The shear forces produced by the fixed percentage method are based on a percentage of the dead weight and the effective mass of liquid. The seismic moment is equal to the shear force multiplied by the distance from the tank base to the effective center of mass. Condition 2 should be used in the event the purchaser does not specify the condition of seismic loading. Unless otherwise specified, the fixed percentage method shall be equal to 2.5, 5, and 10% for tanks located in seismic zones 1, 2, and 3, respectively. For seismic zone 4 and when specified by the purchaser for seismic zones 1, 2, or 3, the moment produced at the base of the tank is determined by the pseudodynamic method outlined in Section 13 of AWWA D100-84. Section 13 further classifies tanks into three categories: cross-braced elevated, pedestal-type elevated, and flat-bottom tanks. For flat-bottom tanks, the method of approach known as effective mass method is based on the same equations outlined in Appendix E of API 650. The approach is the same as API 650, but the moment equation at the tank base and the variables used are different. The primary difference is the addition of a variable K to the moment equation based on the type of tank, which divides flat-bottom tanks into two classes, anchored and unanchored. Typically, the procedure outlined in AWWA generates a moment greater than that obtained using the equations outlined in Appendix E of API 650. From the seismic moment produced, the tank overturning and shell buckling stabilities must be checked. If the tank is unstable against overturning, its design must be changed or anchorage added. The axial stresses produced in the shell from the moment must be checked to guard against buckling. The AWWA allows for an increase in the shell stress as a result of pressure from the product, which is another difference between AWWA and the procedure of API 650.

Section 11 of AWWA D100 covers inspection and testing requirements of a tank. For the inspection of butt-welded shell joints, the standard allows three methods: radiograph examination, removal of sectional segments, and air carbon arc gouging. Of the three tests, AWWA D100-84 states that radiograph examination of seams is the preferred method of inspection. Prior to painting and post inspection, a tank must be tested per Section 11.11 of AWWA D100. For flat-bottom tanks, the lap-welded seams are generally vacuum-tested by the same procedure described for API 650 tanks. If applicable, the shell, the bottom, and the roof below the overflow should be tested for leaks by filling the tank with water. Painting and disinfecting the tank can be completed upon successful inspection and testing.

Appendix C Tank Design

With the advent of high-strength steels, in 1965 the AWWA added Appendix C for the design of tank shells, to be utilized only when specified by the purchaser. The merit of Appendix C is in its higher allowable stress for material that the main body of the standard restricts to 15,000 psi multiplied by a joint efficiency. An example of the higher allowable stress is a comparison between the two parts of the standard for material ASTM A36. Per Table C.4 of Appendix C, A36 has an allowable stress of 19,330 psi, and in the main body of the code it has an allowable stress of 12,750 psi (allowable stress of 15,000 psi times a joint efficiency of 85%). This increase in allowable stress reduces the required shell thickness and subsequently the shell material cost.

The increase in allowable stress leads to an important aspect of Appendix C, which is to ensure that the material has adequate toughness to prevent brittle fracture failure. Appendix C divides the allowed material into the following three categories: category 1, low-strength material listed in Table C.1; category 2, medium-strength material listed in Table C.2; category 3, high-strength ASTM A517 material. The material used is based upon the Design Metal Temperature (DMT) as specified by the purchaser or under special conditions a temperature equal to 15°F plus the lowest one-day mean temperature given in Figure C.3. Category 1 material selection is based on the thickness and the specified DMT given in Table C.1 and is exempt from any impact test requirements. For example, ASTM A36 is allowed as a plate material if the thickness is less than or equal to 0.5 inch when the DMT is equal to or greater than 20°F. Category 2 material is divided into three groups based on the type of heat treatment conducted on the plate. Table C.2 lists the maximum thickness allowed at various DMTs with the maximum thickness for each group equal to 1.5 inches without conducting impact tests. The only material in category 3 is ASTM A517, and impact testing is required for any combination of thickness and DMT.

After material selection, the shell thickness can be based on two different plate thickness equations: the basic AWWA D100-84 equation given by equation (10-1) and the API 650 variable-point method that is only allowed when the height-to-diameter ratio is 0.5 or less.

For both cases, the allowable stress is given in Table C.4, and the joint efficiency is taken as 1.0. As stated in Appendix C, the higher allowable stresses will produce thinner shell plate; so checking the shell stability against buckling is important to determine the need for stiffener rings.

The AWWA publishes a number of other standards for the construction of ASTs and related equipment. Of these standards, the two most relevant are AWWA D-102, previously titled *Standard for Painting Steel Water-Storage Tanks,* which is the standard for painting of tanks, and AWWA D-105, *Standard for Disinfection of Water Storage Facilities*. AWWA D-102 has been pulled from the AWWA publication list, and is currently being rewritten (with a 1995 anticipated publication date) because the previous version specified five different paint systems, four of which are no longer being manufactured due to their lead or high VOC (volatile organic compounds) content. AWWA D-105 describes the requirements for disinfecting a storage tank where the primary function is to remove bacteria. According to the standard, liquid chorine, sodium hypochlorite, or calcium hypochlorite is combined with water at significant concentrations to disinfect the tank's surfaces. The concentrations vary according to the type of application. The disinfectant solution can be applied by spraying the entire surface below the overflow elevation, or the tank can be filled with water to the overflow elevation and disinfectant added. The storage tanks should be properly tested for free chlorine residual and bacteria content following disinfection procedures.

American Society of Mechanical Engineers Pressure Vessels

When the internal design pressure of aboveground storage tanks exceeds 15 psig, ASME Codes govern the design of these tanks, which are referred to as pressure vessels. Because of the significant design pressure of ASME vessels, the requirements are far different from those of any of the other design standards previously discussed. Although the focus of this chapter is on field-erected tanks, the ASME pressure vessels can be very small, such as shells constructed out of pipe material, and they are commonly used for shop-built vessels or components.

Table 10-2. ASME Boiler and Pressure Vessel Code sections

Section	Title
I	Power Boilers
II	Materials
Part IIA	Ferrous Material Specifications
Part IIB	Nonferrous Material Specifications
Part IIC	Specifications for Welding Rods, Electrodes, and Filler Metals
Part IID	Properties
III	Nuclear Power Plant Components
IV	Heating Boilers
V	Nondestructive Examinations
VI	Recommended Rules for Care and Operation of Heating Boilers
VII	Recommended Rules for Care of Power Boilers
VIII	Rules for Construction of Pressure Vessels
Division 1	Pressure Vessels
Division 2	Alternate Rules
IX	Welding and Brazing Qualifications
X	Fiberglass-Reinforced Plastic Pressure Vessels
XI	Rules for Inservice Inspection of Nuclear Power Plant Components

Derived from: ASME.

The ASME Codes standardized the various state requirements of pressure vessels and simplified the procedure of supplying a vessel for both the manufacturer and purchaser. Although the ASME Codes have existed for a number of years, a number of states and local governments still produce requirements for the design of pressure vessels. Therefore, before purchasing a tank, a purchaser must consult local authorities to determine the governing pressure vessel code and/or regulations.

The most current copy of the ASME Code on pressure vessels, *Boiler and Pressure Vessel Code,* is the 1992 edition, along with the 1992 and 1993 addenda. This ASME Code is revised every three years with the next edition set to be published in 1995. In between each new edition of the code, ASME issues addenda published at the end of every year. The ASME Code requirements are divided into 11 sections that govern different aspects of boiler and pressure vessels, as listed in Table 10-2. Section II of the code describes the material allowable stresses. The 1992 edition of the code moved all the material requirements from the different sections into Section II to achieve uniformity of the material requirements of the different code sections. There are two divisions of Section VIII, Divisions 1 and 2, which are pertinent to the design of pressure vessels constructed of steel. Division 1 gives the code rules for basic pressure vessels, and Division 2 outlines alternate rules for the design of pressure vessels. Section V describes the requirements of nondestructive examination, and Section IX lists the welding qualification requirements.

The majority of ASME pressure vessels used in the petroleum industry are vessels constructed according to Section VIII, Division 1. For the remainder of this discussion, "ASME Code" will be used in reference to ASME Section VIII, Division 1 requirements for pressure vessels. Division 1 has three subsections, along with mandatory and nonmandatory appendixes. Subsection A contains Part UG, listing the general requirements for design and construction of pressure vessels. Subsection B consists of three Parts, UW, UF, and UB, which give requirements for vessels fabricated by welding, forging, and brazing, respectively. Subsection C pertains to the requirements as related to the material, the most prevalent being Part UCS, which is the requirement for vessels constructed of carbon and low-alloy steels. Parts UNF and UHA list the material requirements of nonferrous and high-alloy materials, respectively. The current code consists of 26 mandatory appendixes, which cover supplementary rules or clarification of rules on specific applications of the ASME Code. If applicable, these mandatory appendixes are required in the design and the construction of pressure vessels. Nonmandatory appendixes are not required for pressure vessels and are comprised of suggested good practices or give the foundation for some of the ASME Code rules.

A manufacturer of pressure vessels must have the proper authority to construct any vessel according to the ASME Code. Where API and AWWA standards exist as recommended guidelines, the ASME Code offers strict requirements of pressure vessel design. A vessel manufacturer must have an agreement with an inspection agency and apply for a certificate of authorization from the Boiler and Pressure Vessel Committee of the ASME Society. According to Section UG-2 of Division 1, a manufacturer is responsible for meeting the requirements of the code and shall have in place a quality control system that establishes requirements from the initial engineering to the final examination. An authorized inspector must ensure proper quality, verify code conformance, and authorize application of the manufacturer's code symbol. Section UG-117 lists the requirements for obtaining the ASME Code stamp. The inspector must grant the manufacturer approval to affix the code stamp, with the code symbol

FIGURE 10-1.
ASME nameplate stamp.

embossed on the nameplate. Figure 10-1 shows the general code symbol "U" for a pressure vessel, with the "W" below the symbol signifying welded construction. The "RT4" symbol designates the amount of radiograph examination that was conducted. The presence of a code stamp indicates that an authorized inspection agency has certified that the vessel is constructed according to the requirements of the indicated code.

Because the ASME Code allows for the design of pressure vessels up to 3000 psig, the rules of Division 1 are very detailed and complex. Therefore, the authors recommend the following books, which simplify the requirements of ASME: *Pressure Vessels: The ASME Code Simplified,*[1] and *Pressure Vessels Design Handbook.*[2] These books not only simplify the requirements of ASME, but also provide examples of the application of ASME Code rules.

Section UG-16 of the ASME Code begins the discussion of the design of pressure vessels, listing requirements for the design temperature (Section UG-20), pressure (Section UG-21), loadings on the tank (Section UG-21), and allowable stresses (Section UG-23). The design pressure shall be equal to the maximum operating pressure of the vessel. To ensure that vessels are not overstressed, a pressure relief valve should be installed, and the vessel should be designed to account for the overpressure required for proper vent operation. Section UG-134 states rules on the pressure setting of relief valves. The ASME Code requires consideration of the following loads:

- Internal and external design pressure.
- Weight of vessel and contents under operating and test conditions.
- Static and dynamic loads from equipment, piping, and insulation.
- Static reactions from attachments of internals and external supports.
- Cyclic and dynamic reactions from pressure or thermal variations.
- Environment loads such as wind, snow, and seismic loads.
- Impact loads due to fluid shock.
- Temperature gradients and differential thermal expansion.

The material allowable stress under tension is defined in Section II, Part D of the ASME Code for the various materials. The allowable stress for materials under compression is defined in Section UG-23(b) of Section VIII, Division 1. The rules of this section are dependent on the material's buckling strength and are a function of the ratio of the tank radius to the plate thickness and of the modulus of elasticity of the material at the given design temperature. Section UG-23 allows for an increase in allowable stresses when a combination of loads occurs in the plate. The allowable stresses for plate material when the stresses in the plate are a result of primary membrane and bending stresses should not be greater than 1.5 times the allowable tension stress of the plate. The ASME Code does allow for an increase in any allowable stress by a factor of 1.2 when the stress results from loads in combination with wind or seismic loads.

The big difference between atmospheric or low-pressure tanks and pressure vessels is the presence in the latter of high longitudinal as well as hoop stresses. The high pressure and sometimes relatively small diameter also can result in thick plate material. Therefore, the thickness of pressure vessel walls depends on whether the radius is measured from the inside or the outside, and ASME designates sets of equations for both situations. Table 10-3 outlines the plate thickness requirements for cylindrical and spherical shell plate due to internal pressure as taken from Section UG-27 and using the internal radius.

The joint efficiency for determining the shell plate thickness is given in Part UW, which also describes the requirements for welded

Table 10-3. ASME Section VIII shell plate thickness equations

Stresses	Equation	Limits
Cylindrical shell circumferential	$t = \frac{PR_i}{SE-0.6P}$	$P < 0.385SE$
Cylindrical shell longitudinal	$t = \frac{PR_i}{2SE+0.4P}$	$P < 1.25SE$
Spherical	$t = \frac{PR_i}{2SE-0.2P}$	$P < 0.665SE$

Derived from: ASME Section VIII.

t = minimum thickness, inches; P = internal pressure, psi; Ri = inside radius, inches; S = maximum allowable stress, psi; E = joint efficiency.

joints. The section details weld joint efficiency by first classifying the type of joint. The ASME Code defines a weld joint according to its location, putting weld joints into four categories. Categories A and B define longitudinal and circumferential weld joints, respectively. Category C defines welds attaching flanges to nozzles, and in Category D weld joints attach nozzles to a vessel. For an ASME vessel, Table UW-12 states the limitations of weld categories, depending on the type of weld. Double butt welded joints are allowed for any weld category, but single fillet welds have restrictions based on the application and generally are not allowed for longitudinal or circumferential weld seams. For full-penetration butt welds, the ASME Code allows a joint efficiency of 1.0 for full radiograph, 0.85 for spot radiograph, or 0.70 for no radiograph examination. Analogous to API 620, ASME requires that seams with a joint efficiency of 1.0 be fully-radiographed. Section UW-12 has additional requirements for reduced joint efficiency and allowable stress when the shell or heads are seamless.

An ASME vessel typically is designed not only for internal pressure but commonly for external pressure as well. The gauge of external pressure or vacuum can be as great as 15 psig or full vacuum. The ASME rules for pressure vessel design to withstand vacuum loads are given in Section UG-28 for shells and based on the maximum spacing between stiffeners. Vacuum rings installed on pressure vessels are analogous to wind girders on flat-bottom tanks, which maintain shell stability under wind loads. The spacing of vacuum rings is a function of the shell thickness, with thinner shell plate requiring more frequent rings than thicker shell plate. Therefore, the spacing of vacuum rings should be varied with the shell plate thickness to produce the most economical shell design.

PHOTO 10-1.
ASME pressure vessell spheres.

Aboveground storage tanks designed with an internal pressure near atmospheric or under low pressure can possess anchors to adequately prevent uplift of the tank shell. For high-pressure vessels, it is impractical to anchor flat-bottom tanks, and this is not allowed per the ASME code. Therefore, the vessels can be supported in either the vertical or the horizontal position. Vertical tanks are supported on a skirt, support legs, or support lugs or rings, whereas horizontal vessels are supported on saddles. Skirts for vertical vessels are simply shell sections that support the pressure vessel. Externally, the bottom head of a pressure vessel is not visible because of the skirt. Support legs are structural members such as wide flange beams or pipes that are attached to the external vessel surface for support. Support legs usually are cross-braced for lateral loads, the same type of support used on older elevated water tanks or silos. Support lugs or rings are attached to the vessel wall and usually are used to support vessels by a structural framework. Saddles are used almost exclusively to support horizontal pressure vessels which commonly are referred to as bullets. Typically two saddles are located on the shell, with the distance from the shell to the head seam equal to about 20% of the shell length. One saddle typically is fixed, and the other slides to allow movement of the vessel due to temperature fluctuations. The use of saddles is the most economical way of supporting a pressure vessel, but this also requires the most land space of any support system.

Notes

1. *Pressure Vessels: The ASME Code Simplified,* fifth edition, Robert Chuse, McGraw-Hill, New York, NY, 1977.
2. *Pressure Vessels Design Handbook,* Henry H. Bednar, PE, Van Nostrand Reinhold, New York, NY, 1981.

CHAPTER 11

Underwriters Laboratories Codes

The main standard-setting body for shop-built ASTs in the United States is Underwriters Laboratories Inc. (UL). Founded in 1894, UL is an independent, not-for-profit organization that operates laboratories for the examination and testing of products for public safety. In regard to ASTs, Underwriters Laboratories has two codes that will be examined here: UL 142, *Standard for Steel Aboveground Tanks for Flammable and Combustible Liquids,* Seventh Edition and UL 2085, *Insulated Aboveground Tanks for Flammable and Combustible Liquids,* First Edition.

Underwriters Laboratories Standard 142 ////////

UL 142, *Standard for Steel Aboveground Tanks for Flammable and Combustible Liquids,* Seventh Edition, became effective in April,

1994. This standard covers construction methods for steel, atmospheric shop-built ASTs, addressing primary tanks, secondary tanks, diked tanks, tank supports, tank accessories, and test methods. UL 142 ASTs are shipped from the manufacturer as completely assembled vessels bearing the UL label. However, UL 142 does not apply to other ASTs covered by API 650 Appendix J or by API 12F.

Tank Construction

In manufacturing a tank, only new material of commercial or structural grade carbon steel or stainless steel shall be used. Carbon steel shall comply with ASTM A36, A569, or A635, or have a concentration of carbon that is less than 0.3%. Stainless steel shall comply with ASTM A167 or A240.

Wherever two pieces of steel plate are welded together in constructing a tank, that area is referred to as a joint. UL 142 gives manufacturers a range of options in joint construction. There are various joint types, including shell, head, bottom, roof, and corner joints. Some manufacturers offer the customer a choice of joint construction methods, but one should be aware that certain methods are more costly than others. For example, a simple butt weld joint is more cost-effective and easier to make than a welded joint utilizing offset construction. To visualize various joint attachments, Figures 11-1 through 11-5 should be consulted:

- Figure 11-1 shows a few samples of joints formed by attaching two pieces of steel plate together.
- Figure 11-2 shows various options for attaching the head of a tank to the shell on a horizontal cylindrical tank.
- Figure 11-3 shows all the options for attaching the tank bottom to the tank shell for a vertical cylindrical tank.
- Figure 11-4 shows examples of methods used to attach steel plate roofs to the tank shell.
- Figure 11-5 shows all the options for making corner attachments on rectangular or dike tanks.

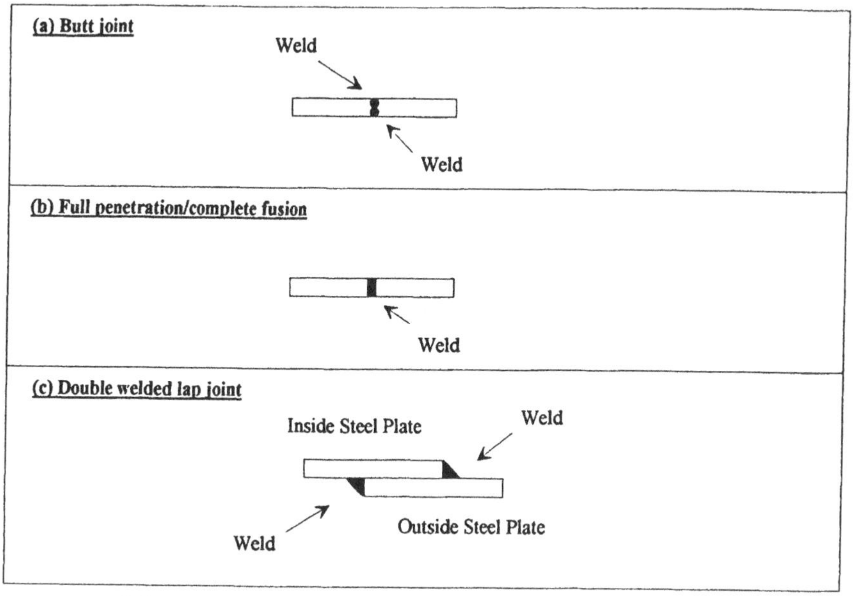

FIGURE 11-1.
Sample shell joints. (Source UL 142)

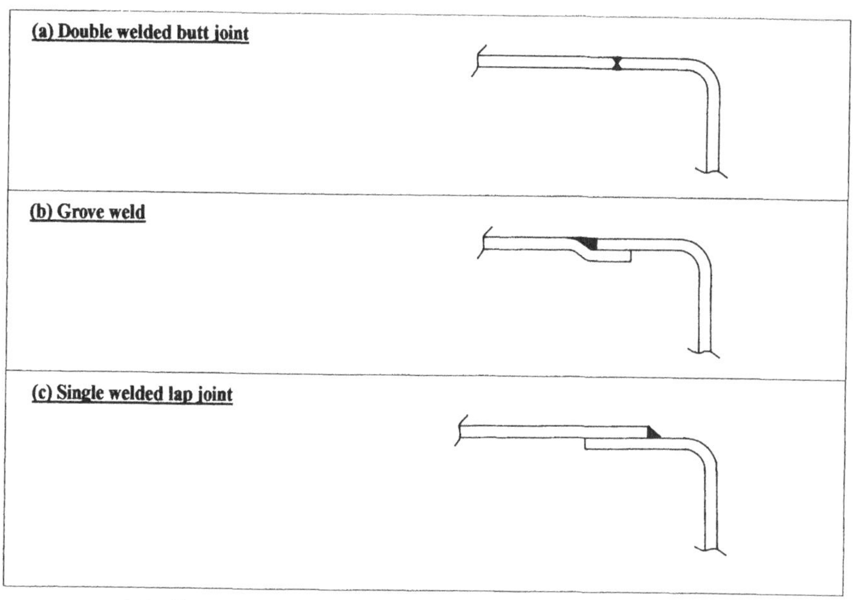

FIGURE 11-2.
Sample head joints. (Source UL 142)

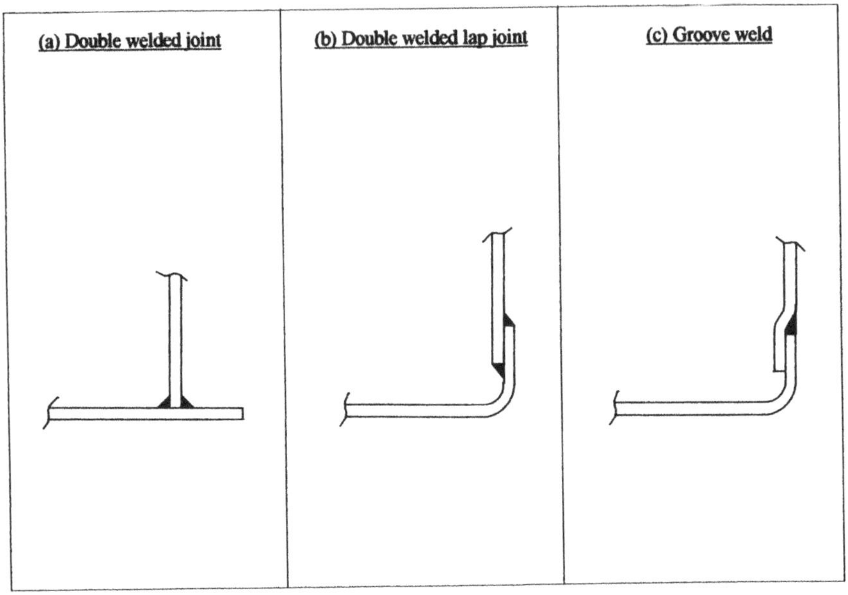

FIGURE 11-3.
Sample bottom joints. (Source UL 142)

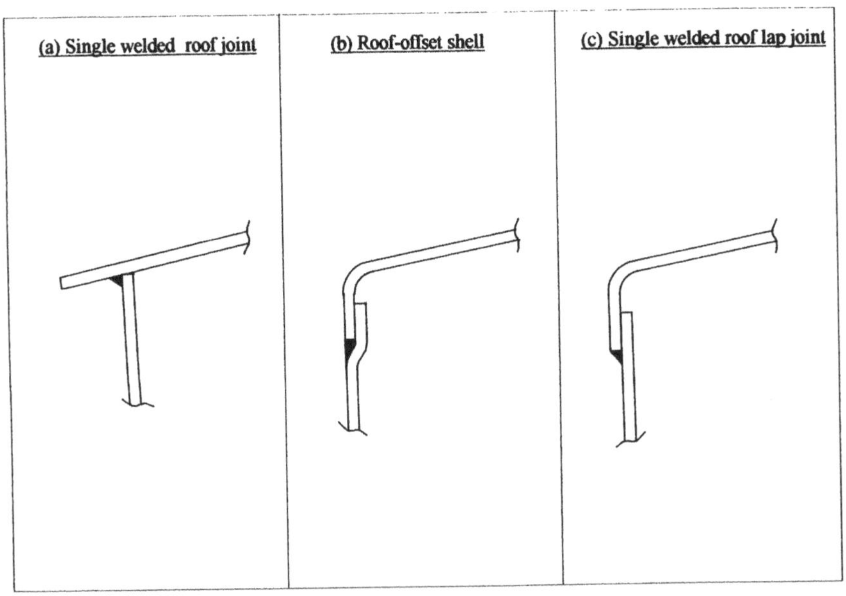

FIGURE 11-4.
Sample roof joints. (Source UL 142)

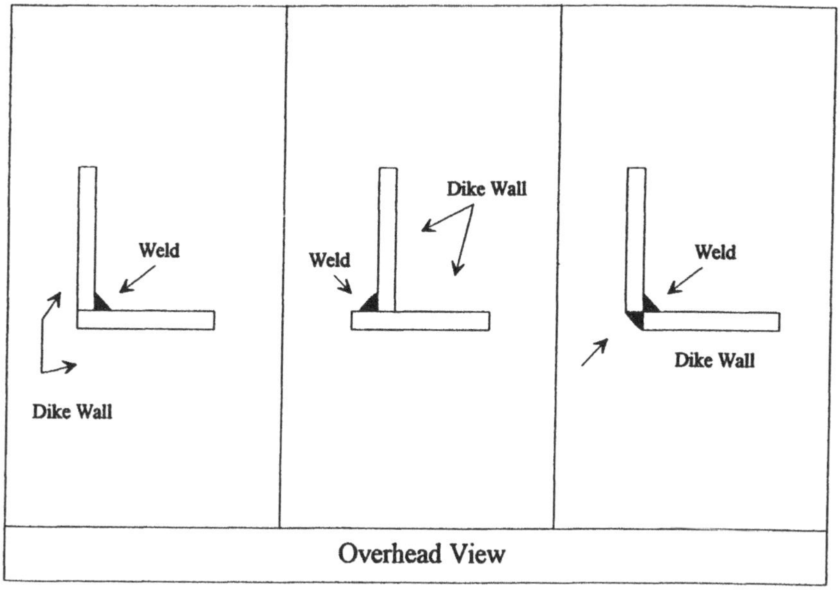

FIGURE 11-5.
Sample corner joints (rectangular and dike tanks). (Source UL 142)

Tank connections are to be provided for each tank opening by welding either a steel pipe coupling, a threaded steel flange, or a standard pipe nipple to the tank. Alternatively, a steel flange welded to a length of pipe that is welded to the tank may be used. All pipe connection openings shall be protected with metal or plastic plugs when the tank is stored or during transit. Figure 11-6 shows examples of a threaded coupling and a standard pipe nipple.

Venting is of major concern in the safe operation of ASTs, so UL 142 covers the basic venting requirements for all tanks. Each primary tank or each compartment of a multi-compartment tank shall have both normal and emergency venting. In addition, the interstitial space of a secondary containment (double wall) tank also shall have emergency venting. In general, emergency vents for shop-built ASTs range in size from two inches to ten inches.

There are a few options for tank manufacturers in providing emergency venting. Most likely, an emergency vent will be attached to the pipe connection. On vertical cylindrical tanks, emergency venting may be accomplished by employing a weak shell-to-roof

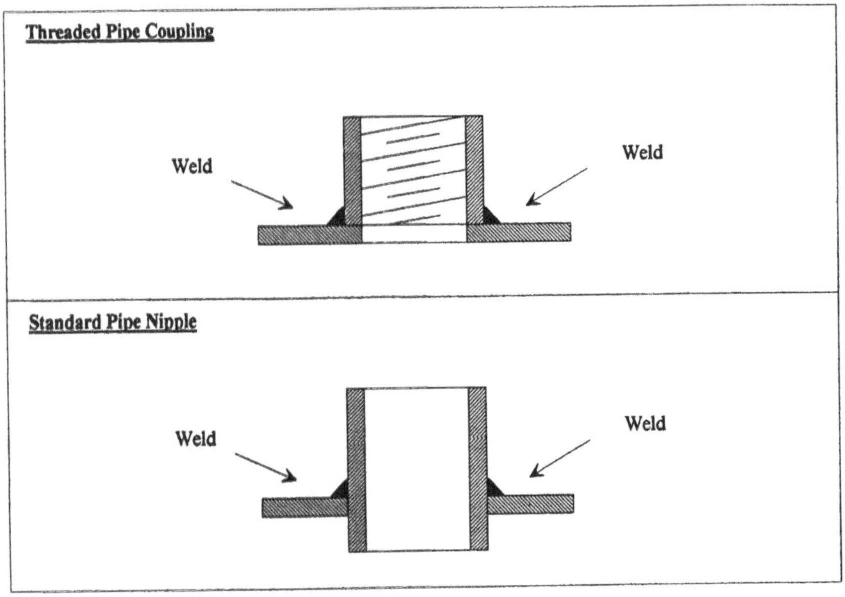

FIGURE 11-6.
Threaded coupling/standard pipe nipple. (Source UL 142)

joint construction. Additionally, on both horizontal and vertical ASTs, a long-bolt manway or a manhole on top of the tank with a cover fabricated to lift under an internal pressure of 2.5 psig can be used for emergency venting.

Other openings for ASTs, such as manways and fill openings, are required by UL 142. A manway shall be included in each tank or compartment of a multi-compartment tank that has over a 5000-gallon capacity. Fill and gauge openings in the top of the tank must be provided to connect fill piping and to install a liquid level indicator. All openings on the top of the tank shall be above the highest liquid level. UL 142 allows an opening in the bottom of the head or sidewall for drainage of the primary tank, of the interstitial space of a secondary containment tank, or of a dike wall.

Painting of the AST is required by UL 142, unless stainless steel is used, to protect the tank from external corrosion during storage at the factory or during transit. This can be accomplished by applying one coat of paint. Most manufacturers will apply one coat of shop primer to the exterior of the tank to satisfy this requirement.

Construction of Primary Tanks

Horizontal Cylindrical ASTs

Horizontal aboveground storage tanks have capacity and diameter limits per UL 142. Furthermore, the length of a horizontal AST is limited to six times its corresponding diameter. Refer to Table 11-1 for tank capacity as compared to tank diameter specifications.

Table 11-1. Capacity vs. maximum diameter—horizontal tanks

Actual capacity (gallons)	Maximum diameter (inches)
550 or less	48
551–1,000	64
1,001–9,000	76
9,001–35,000	144
35,001–50,000	144

Source: Developed from UL 142.

In constructing horizontal tank heads, UL 142 states that the head shall be constructed of not more than three sections for tank diameters of 48 to 96 inches or four sections for tank diameter of 97 to 144 inches. Also, it is possible to use the standard flat head, or, alternately, a dished head is allowable on a horizontal tank. UL 142 lists minimum depth measurements for dished heads as compared to tank diameter.

Compartment Tanks

The main concern in construction of a multi-compartment tank is the design of the bulkhead between compartments. UL 142 specifies that a bulkhead must be constructed so as not to allow any leakage from one bulkhead joint to flow into the adjoining compartment. This can be accomplished as shown in Figure 11-7. All single bulkheads are to be constructed of one piece of material, which may be either flat or dished. Conversely, double bulkheads with tank diameters of 48 to 96 inches shall be constructed of not more than three pieces, and double bulkheads with tank diameters of 97 to 144 inches shall be constructed of not more than four pieces.

Vertical Cylindrical ASTs

There are some specific guidelines that UL 142 requires in the construction of vertical cylindrical ASTs. The maximum capacity for

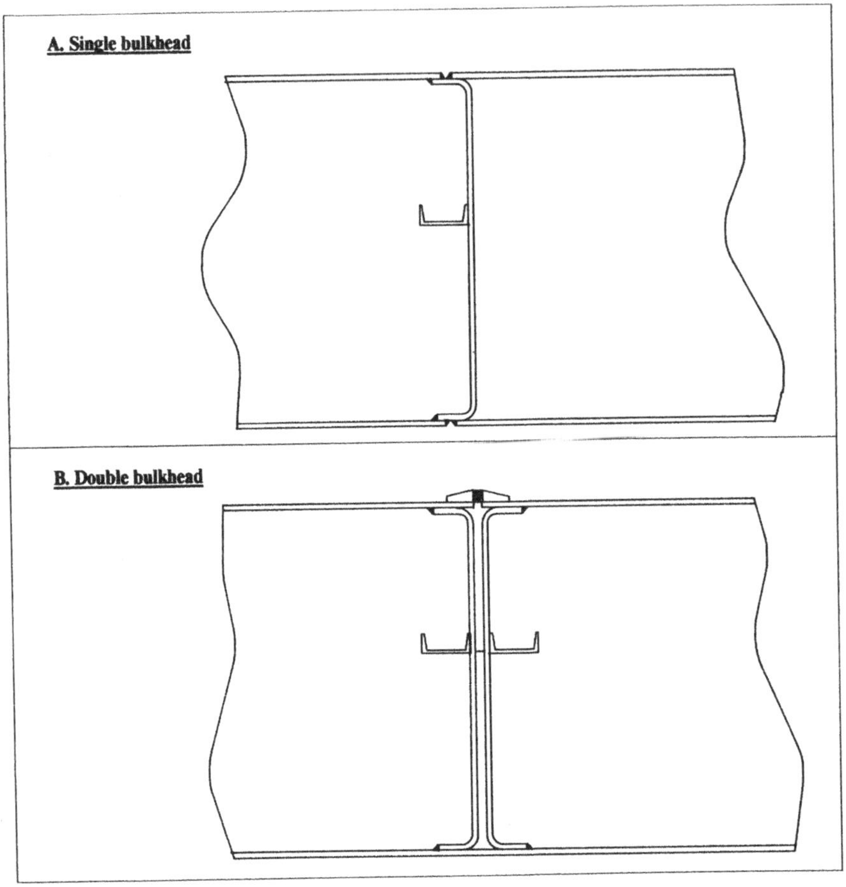

FIGURE 11-7.
Bulkheads for compartment tanks. (Source: UL 142) .

a UL vertical tank is 50,000 gallons, and the height of the tank cannot exceed 35 feet. In addition, the minimum diameter of the vertical tank must be at least one-quarter of the height of the tank. (*Example:* A 20-foot-high vertical tank would have to be at least 5 feet in diameter.) In constructing the top or the roof of a vertical AST, no more than four pieces of material may be used, and the roof should be either dished or conical. (*Exception:* A flat-top roof is acceptable if it can withstand the UL 142 top load test and tank leakage test.) Many manufacturers use weak shell-to-roof joint construction techniques as a form of emergency venting and safety. The purpose of weak shell-to-roof joint construction is that under extreme internal pressure, the frangible joint connecting the roof to

PHOTO 11-1.
Single wall UL 142 ASTs installed at a remote cardlock facility.

the shell will give way and release the pressure buildup in the tank before the shell joints rupture. If the vertical tank is constructed with a weak shell-to-roof joint, then the roof must be conical with a minimum tank diameter of ten feet. Also, on vertical tanks, the bottom or the floor shall be constructed of no more than four pieces of steel. (An example of this type of AST is shown in Photo 11-1.)

Rectangular Tanks

UL 142, for the first time, addresses rectangular primary containment tank construction in this edition of the code. This section allows for the use of stiffening bars or tie rods in the construction of the rectangular tank. In order to meet the requirements of this section, the rectangular primary containment tank must pass both the UL top load test and the UL leakage test.

Construction of Secondary Containment Tanks

UL makes a point of definition in this section by specifying that all secondary containment tanks must be provided with a means to monitor the interstitial space of a secondary containment tank for leakage. This is to be accomplished by allowing any liquid that enters the interstitial space to flow to the point of monitoring. (*Example:* A

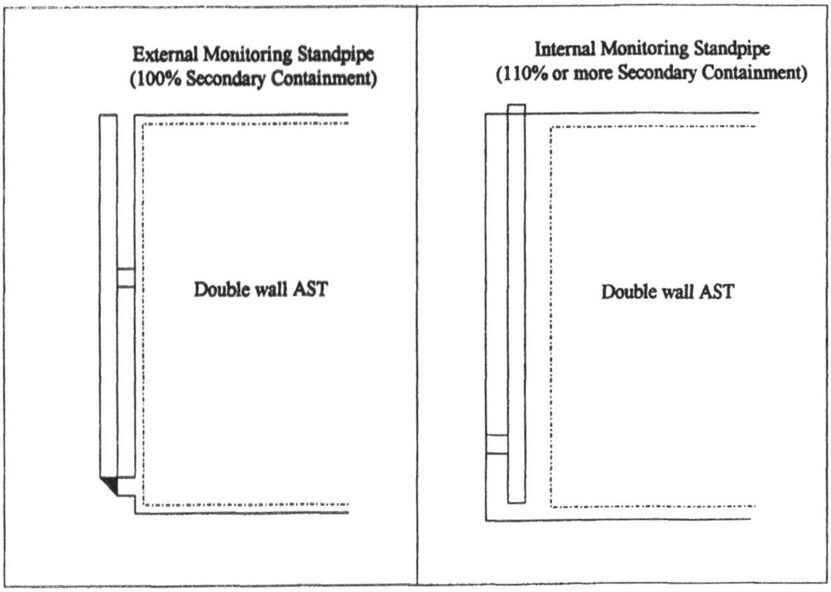

FIGURE 11-8.
Secondary tank designs (Type I and Type II).

pipe is installed on either the internal or the external wall of one end of the tank to act as a monitoring well for the interstitial space.)

Horizontal Secondary Containment Tanks

For a horizontal secondary containment tank, construction of the primary tank and of the outer or secondary tank follows the standard requirements of UL 142. Manufacturers have a choice of the distance between the two tanks. The industry standard addresses two kinds of horizontal secondary containment tanks: Type I or Type II. A Type I design will have the primary and the secondary tank in intimate (direct) contact with each other with a monitoring pipe on one exterior end of the tank. Conversely, a Type II design will have an extended gap on one end of the secondary tank, and the monitoring pipe can be internally located in the interstitial space. Figure 11-8 illustrates the differences between Type I and Type II configurations.

Vertical Secondary Containment Tanks

Vertical secondary containment tanks are to be constructed in the same manner as the standard UL 142 vertical primary tanks. The tank floor construction must incorporate two separate and distinct tank bottoms. Most manufacturers use a type of standoff or separation material between primary and secondary bottoms to ensure proper monitoring flow. Unlike primary vertical tanks, secondary vertical tank roofs may not use a weak shell-to-roof joint construction for emergency venting.

Also, UL 142 covers rectangular secondary containment tank construction. The main point is that the floors of the primary and the secondary tanks must be separate, and the rectangular tank must pass the UL leak test and the UL load test.

Diked Tanks

UL 142 discusses construction details of open- and closed-top dike tanks in this section. Unlike secondary containment tanks that only require 100% containment of the primary tank, a steel dike must contain 110% of the primary tank capacity. Also, horizontal cylindrical tanks must be provided with supports that are secured to the tank, the dike, or both. Additionally, the supports must allow the free flow of liquid throughout the diked area. In an open-top design, if the dike wall is in excess of six feet, a ladder or stairs shall be provided for entry into the diked area.

For closed-top dike designs, a steel cover will be attached over the diked area to prevent debris from entering the area and allowing for its inspection. UL states that the inside of the diked area shall be designed so that it cannot be pressurized, and it must have a means of emergency venting. All vent openings, in this design, should guide vapors upward. Finally, for both open- and closed-top designs, UL requires the systems to pass the UL leak test and the UL buoyancy test.

Tank Supports and Accessories

Tank supports are engineered to resist a load equal to two times the weight of the full tank. The saddles shall have a maximum height of 12 inches unless they are protected by a fire-resistant material with a two-hour fire rating. (To avoid this rule, most manufacturers build the saddles to a height of 9 to 11 inches.) Furthermore, UL has guidelines on the length of the saddle base plate (at least 90% of the tank diameter) and the proper positioning of the saddles on the tank (the tank diameter divided by four). For example, an eight-foot-diameter tank would have saddles located two feet from each end and a base plate length of at least 7′2″). Refer to Figure 11-9 for an

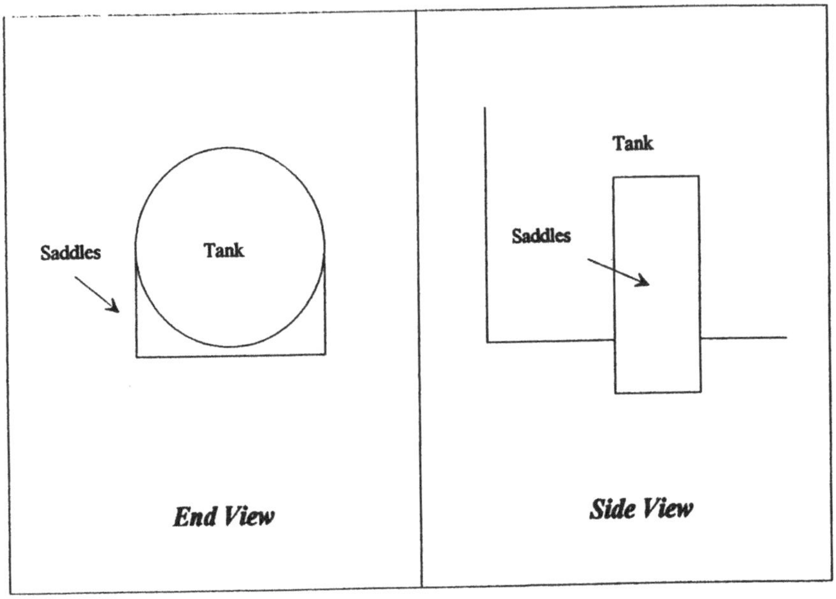

FIGURE 11-9.
Steel saddles.

example of steel saddles. For tank accessories, UL 142 requires that all ladders, stairs, and runways comply with the appropriate current edition of OSHA.

Performance Test Methods

UL 142 discusses these performance test methods: the tank leakage test, the hydrostatic strength test, the top load test, the buoyancy test, the hydrostatic load test, and the tank support load test.

The tank leakage test for primary tanks is accomplished by using one of the following methods (*note:* each compartment of a tank having two or more components shall be tested separately):

- Using internal pressure and a soap-based solution applied to the seams and joints of the tank to check for leaks. Horizontal or rectangular tanks shall use an air test pressure between 3.0 psig and 5.0 psig. Vertical tanks shall use an air test pressure between 1.5 psig and 2.5 psig.
- Filling the tank with water and applying 5.0 psig hydrostatic test pressure, and then examining the tank seams and joints for leakage.

The tank leakage test for secondary tanks has some additional steps. First, once the primary tank has been manufactured, it shall be tested as described above. Then, after the secondary tank has been attached around the primary tank, the primary tank shall be air-pressurized again, and the air pressure will be held for one hour. A drop in pressure would indicate a leak. Once the one-hour test is completed, the interstitial space shall be pressurized as above with the pressure maintained on the primary tank. A soap-based solution applied to the seams and joints will be used to check for any leaks. Alternatively, a secondary containment tank may be tested by applying to the interstitial space a vacuum of at least 13 inches of mercury for 12 hours. If the tank can maintain the mercury level (±2 inches of mercury) for the required time, the test is successful.

The hydrostatic strength test calls for the primary tank to be filled with water and then for pressure to be applied to the tank in increments of 5.0 psig at a maximum rate of 2.0 psig per minute until the pressure reaches the desired rate of 25.0 psig. At 25.0 psig, the primary tank shall be examined for leakage or rupture. To test the secondary tank, the interstitial space shall be filled with water, with the same procedure then followed for testing the primary tank while maintaining pressure in the primary tank.

The top load test will test the flat top surface of a tank by placing a 1000-pound load over a one-square-foot area for a period of five minutes. Then, after removal of the load, the tank will be subjected to tank leakage test.

The buoyancy test requires that the entire diked area be filled to capacity with water while the tank is completely empty for a period of one hour. The tank must not lift from the dike floor to pass this test. A visual inspection of the dike and the tank for any damage shall be conducted once the dike is emptied.

The hydrostatic load test is to be conducted with the buoyancy test. When the dike is empty, the primary tank is completely filled with water, and this condition is held for a period of one hour while the tank is checked for leaks.

The tank support load test is done to test the integrity of the tank supports. To accomplish this test, the tank is filled with water, and an evenly distributed load, equal in weight to the filled tank, is applied to the top longitudinal axis of the tank for a period of two minutes.

Tank Markings

As the tank is completed, various safety markings shall be applied to its exterior to allow operating personnel to handle the tank safely once it leaves the factory. Compartment tanks should have the appropriate markings for each compartment. UL 142 requires that each tank that leaves the factory under the UL label have the following markings:

1. Manufacturer's name or trade name and factory location, if the company has multiple locations.
2. Identification of emergency vent openings.
3. The statement: "This tank is intended for stationary installation only."
4. One of the following, as applicable:
   (a) "This tank requires emergency relief venting."
   (b) "Built to weak shell-to-roof joint design."
   (c) "This manhole is provided with long bolts to permit emergency relief venting."

Secondary containment tanks shall have the following additional markings:

1. "The annular space requires emergency relief venting."
2. "Pressurize primary tank when pressure testing annular space."

For diked tanks, the capacity in gallons of the diked area shall be clearly identified.

Underwriters Laboratories Standard 2085

UL 2085, *Insulated Aboveground Tanks for Flammable and Combustible Liquids,* First Edition, was finalized in 1994. This subject covers insulated aboveground atmospheric steel tanks storing flammable and combustible liquids. The UL 2085 standard states that a tank is designed as a fire-resistant tank, a protected tank, or both. The difference between the designations has to do with the particular fire code that must be met. A *fire-resistant tank* is intended for use and installation according to NFPA 30 and 30A (refer to Chapter 6).

A *protected tank* is to be used and installed according to UFC Appendix II-F (refer to Chapter 7). Both of these tank types are tested for leaks and shipped from the factory as completely assembled shop-built ASTs.

Construction

A tank built to UL 2085 will have a primary tank and a secondary tank built to the UL 142 standard. The insulation system that protects the primary tank during a fire shall completely encase the primary tank except where a fitting or a tank connection is attached. In addition, the insulation system must not interfere with proper venting or monitoring of the secondary containment. Any openings must be through the top of the tank above liquid level. All UL 2085 tanks shall be supplied with lifting lugs for proper lifting of the tank. The maximum capacities of the tank types are listed in Table 11-2.

Table 11-2. UL 2085 maximum capacities

Tank type	Maximum capacity
Fire-resistant tanks	12,000 gallons
Protected tanks	10,000 gallons

Source: Developed from UL 2085.

Both normal and emergency venting is required on the primary tank, whereas only emergency venting is required of the secondary containment. Also, each compartment of a multi-compartment primary tank must have both a normal and an emergency vent. As in UL 142, a long bolt manhole may be used for emergency venting.

In addition, UL 2085 covers spill and overfill containment even though that is *not* required per this standard. Basically, it states that spill and overfill containment shall be welded to the tank and have a capacity between 5 and 25 gallons. The containment shall have a drain to the primary tank and be equipped with a lockable cover. For overfill prevention device specifications, refer to Table 11-3.

Tank Handling/Installation Tests

Prior to the fire test, UL 2085 requires some additional tank testing techniques, which may be waived under the following conditions:

Table 11-3. Overfill prevention devices

Tank type	Audible alarm sounds at:	Spill prevention device activates at:
Fire-resistant tank	90% of capacity	95% of capacity
Protected tank	85% of capacity	90% of capacity

Source: Developed from UL 2085.

1. *Fire-resistant tanks* with six inches of concrete acting as the insulation material or an external steel wall that protects the insulation material can waive the ball impact test, the cold exposure and impact test, and the simulated environmental exposure and small-scale fire test.
2. *Protected tanks* with six inches of concrete acting as the insulation material can waive the ball impact test, the cold exposure and impact test, and the simulated environmental exposure and small-scale fire test. Also, if the protected tank has an external steel wall that protects the insulation material, the hose stream test can be waived.

Prior to painting the protected AST, the manufacturer shall perform a leakage test, which is almost identical to that of UL 142, to determine the integrity of both the primary and the secondary tanks.

The leakage test for primary tanks is accomplished by using one of the following methods (*note:* each compartment of a tank having two or more compartments shall be tested separately):

- Using internal air pressure and a soap-based solution applied to the seams and joints of the tank to check for leaks. Horizontal or rectangular tanks shall use an air test pressure between 3.0 psig and 5.0 psig. Vertical tanks shall use an air test pressure between 1.5 psig and 2.5 psig.
- Filling the tank with water and applying 5.0 psig of hydrostatic test pressure, and then examining the tank seams and joints for leakage.

The leakage test for secondary tanks has some additional steps. First, once the primary tank has been manufactured, it shall be tested as described above. After the secondary tank has been attached around the primary tank, the primary tank shall be air-pressurized again, and the air pressure will be held for one hour. A drop in pressure would indicate a leak. Once the one-hour test is complete, the interstitial space shall be pressurized as above while pressure is maintained on the primary tank. A soap-based solution applied to the seams and joints will be used to check for any leaks.

Another testing option for secondary tanks is to pull a vacuum on the interstitial space. The vacuum shall be at least 13 inches of mercury for a 12-hour minimum period. If the secondary tank can maintain the vacuum (±2 inches of mercury), it is a successful test.

The strength of pipe fitting test will ensure that the tank fittings will not crack or split, and that the fitting threads will not strip. This test is accomplished by bending and applying torque to the fitting.

The strength of lift fitting test requires the lifting lugs on the tank to support the empty insulated tank for one minute with a load equal to twice that of the empty insulated tank.

Ball impact tests will test tank integrity by using a four-inch-diameter steel ball having an impact energy of 72 foot pounds. Ball impacts are to be directed at all sides, corners, and fittings of the tank, with no two impacts at the same point. The tank and the insulation system shall be checked for damage.

The load test mandates that a 1,000-pound load be applied to a 12-inch-square surface over various locations on the top of the tank for a period of one minute. The tank and the insulation system shall be checked for damage.

The vehicle impact test requires a 12,000-pound force impacting at ten miles per hour over a one-square-foot area approximately 18 inches above the bottom of the tank. If protective guard posts (vehicle impact barriers) are to be installed at the tank installation, this test may be waived. The intention of UL is to make manufacturers require protective guard posts as part of the tank installation instructions.

The projectile test is an optional test as discussed in UFC Appendix II-F (Chapter 7). The test consists of an empty tank being subjected to five shots of 150 grain M-2 ball ammunition, with a muzzle velocity of 2700 feet per second, fired from a .30 caliber rifle at a distance of 100 feet.

Full-Scale Fire Test

After subjecting the AST to the above-listed tests, it is time to evaluate the thermal resistance of the AST system to a rapid-temperature-rise fire exposure. The full-scale fire test will test the tank size with "the greatest ratio of fire exposed surface area to actual tank volume." To make the test representative of the actual tank assembly in the field, all emergency venting, normal venting, fill pipes, anti-siphon devices, fuel shutoff valves, and similar equipment shall be installed in the tank. All other openings or pipes are to be sealed off.

Thermocouples are to be placed on the interior or exterior of the primary tank and structural supports at a ratio of not less than one thermocouple per tank facing (i.e., head or shell) evenly placed with a maximum distance of nine square feet between the thermocouples. The insulation material of the tank sample to be tested should be allowed to cure properly to manufacturer's specifications, and the test sample should be conditioned to an appropriate moisture condition, as will be achieved in a field installation.

For the AST to pass this test, the following conditions will apply:

1. For a fire-resistant tank, the recorded temperature rise on the primary tank for the two-hour test "shall not exceed an average maximum temperature rise of 800°F," and no single thermocouple may exceed 1000°F. In addition, the venting shall function properly, and the tank structural integrity shall be maintained. Immediately after the fire test, the primary tank shall be subjected to a leakage test (hold 3.0–5.0 psig for one hour).
2. For a protected tank, the recorded temperature rise on the primary tank for the two-hour test "shall not exceed an average maximum temperature rise of 260°F," and no single thermocouple may exceed 400°F. In addition, the venting shall function properly, and the tank structural integrity shall be maintained. Immediately after the fire test, the primary tank shall be subjected to a leakage test (hold 3–5 psig for one hour) and a hose stream test.

Additional Testing

Several additional tests will be made, as follows:

The hose stream test consists of a 30.0-psig hose stream impact from a 2.5-inch-diameter hose applied for a minimum of 2.5 minutes for every 100 square feet of area.

The cold exposure and impact test calls for three test samples to be subjected to cold storage at minus 20°F for 16 hours minimum. Upon removal of the test samples from cold storage, the samples and three additional test samples used for control purposes shall be subjected to an impact test from the 7.08-foot-pound impact of a two-inch-diameter steel ball. The purpose of this test is to determine if the insulation material will crack. If cracking occurs, then the samples shall be subjected to the simulated environmental exposure and small-scale fire test.

The simulated environmental exposure and small-scale fire test is done to test the compatibility of the protective material with the environment. The exposure conditions include: UV light and water, aging, high humidity, industrial atmosphere, salt spray, or combination wet-freeze-dry cycling. A test sample will represent the tank and will be subjected to the selected simulated exposures prior to the small-scale fire test. The small-scale fire test shall last between 50 minutes and 90 minutes and test the integrity of the protective material to fire exposure after it has been subjected to environmental exposure characteristics.

The interstitial communication test is used to check the proper functioning of the monitoring point installed in the annular space of the secondary containment tank, where the insulation material is contained within the interstitial space. First, the primary tank is filled with water. Then, at the opposite end of the tank from the monitoring point, add water, rated vacuum or pressure, to the interstitial space. The substance applied to the interstitial space must be detected through the monitoring point in less than 24 hours to pass the test.

Tank Markings

As the tank is completed, various safety markings shall be applied to its exterior to allow operating personnel to handle the tank safely once it leaves the factory. Compartment tanks should have the appropriate markings for each compartment. UL 2085 requires each tank that leaves the factory under the UL label to have the following markings:

- Manufacturer's name or trade name and factory location, if company has multiple locations.
- Month and year of manufacture.
- Identification of tank connections.
- Tank capacity of the primary tank.
- The statement "Follow Installation Instructions."
- "This tank requires emergency relief venting."

- If applicable, "This manhole is provided with long bolts to permit emergency relief venting."
- "These tanks have not been investigated to determine acceptability for use after fire exposure."
- Tanks complying with the vehicle impact test to be marked "Vehicle Impact Resistant."
- Tanks complying with the projectile test to be marked "Projectile Resistant."
- Markings to be put at the fill fitting describing the required filling procedure.
- Tanks with spill/overfill systems to be marked on the lid "Keep closed and locked when not in use."
- "This tank is intended for installation in accordance with ______" (NFPA 30, 30A, or 31 for fire-resistant tanks, or Uniform Fire Code for protected tanks).

Secondary containment tanks shall have the following additional markings:

- "The annular space requires emergency relief venting."
- "Pressurize primary tank when pressure testing annular space."

CHAPTER 12

Steel Tank Institute Codes

The Steel Tank Institute (STI) was established in 1916 and is a not-for-profit organization consisting of numerous steel tank manufacturers that primarily make shop-built storage tanks and accessories. The larger STI members also fabricate and construct field-erected ASTs. Among the many functions and services that STI performs is the issuance of product standards for shop-built storage tanks and auxiliary equipment. One benefit of an STI-licensed product is a quality assurance program that individual manufacturers must adhere to, including unannounced plant inspections.

This chapter focuses on shop-built AST standards, including: Steel Tank Institute F911-93, *Standard for Diked Aboveground Storage Tanks (DAST)*; Steel Tank Institute F921-93, *Standard for Aboveground Tanks with Integral Secondary Containment;* and Steel Tank Institute F941-94, *Fireguard™ Thermally Insulated Aboveground Storage Tank Standard*. Also, Steel Tank Institute R893-89, *Recom-*

mended Practice for External Corrosion Protection of Shop Fabricated Aboveground Tank Floors, is discussed briefly.

Steel Tank Institute Standard F911-93

Steel Tank Institute F911-93, *Standard for Diked Aboveground Storage Tanks (DAST),* Second Edition, became effective in 1993. The STI F911 standard originally was published in 1991, but, upon UL approval of the system, an updated standard was produced in 1993. The scope of this national standard relates to the construction of steel dikes, which are used as secondary containment surrounding a cylindrical UL 142 primary tank.

The F911 standard requires that the primary storage tank and supports be built to the UL 142 standard, including UL leakage and load tests. The dike size shall contain at a minimum 110% of the primary tank capacity and be able to resist a full hydrostatic load without a rupture. The design of the dike is completely UL-listed and intended for aboveground use only. Refer to Figure 12-1 for an overview of the diked AST elements.

Manufacturers are allowed to fabricate the dike of either carbon or stainless steel. (*Note:* If the dike is made of carbon steel and the tank is constructed of stainless steel, then electrical isolation material should be attached between the tank and the dike.) The main aspects of the F911 standard, including capacity and dimensions, are listed in Table 12-1. The plate thickness given in Table 12-1 is for both the floor and the side walls of the dike. F911-93 states that the "minimum floor thickness of the containment dike shall be at least equal to or greater than the wall thickness, but not less than 10 gauge." Tank support structures can be attached to the dike floor and/or side walls but no higher than a maximum of 12 inches above the dike floor (*reason:* NFPA 30 states that any tank supports above 12 inches must be fire-resistant).

Baffles and/or stiffeners are attached to the inner dike wall to help maintain its structural integrity. Another requirement for the dike tank is the attachment of hold-down straps to keep the tank stationary, to counteract rotational movements or buoyancy. Lifting lugs shall be attached to each tank and the dike side wall, which will allow the tank and the dike to be adequately supported during installation. F911-93 allows a drain opening in the side wall of the

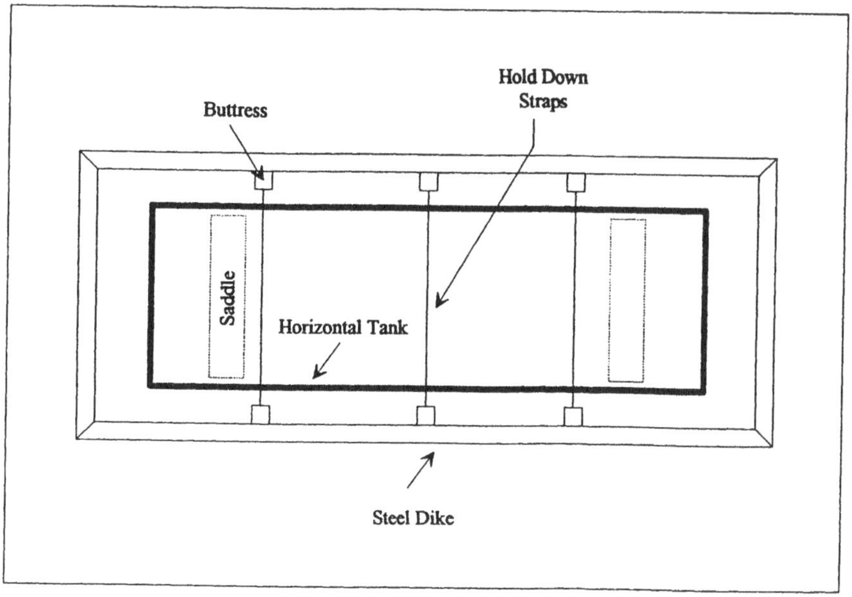

FIGURE 12-1.
Overview of F911 dike.

Table 12-1. Diked AST construction detail

Tank capacity (gallons)	Dike capacity (gallons)	Dike length (max—inches)	Dike width (max.—inches)	Dike height (max.—inches)	Plate thickness (min.—inches)
300	330	84	48	20	10 ga.
550	605	94	72	21	10 ga.
1,000	1,100	108	80	30	3/16
2,000	2,200	180	96	30	3/16
3,000	3,300	240	96	34	3/16
4,000	4,400	324	96	34	3/16
5,000	5,500	204	130	48	1/4
6,000	6,600	248	130	48	1/4
8,000	8,800	332	130	48	1/4
10,000	11,000	420	130	48	1/4
12,000	13,200	444	144	48	1/4
15,000	16,500	444	144	60	1/4
20,000	22,000	590	144	60	1/4

Source: Developed from STI DAST Standard F911-93.

tank for dike-draining purposes, but no opening is allowed in the dike floor.

PHOTO 12-1.
STI-F911 open-top diked AST. (Source: Steel Tank Association)

The manufacturer shall inspect and test each dike for leaks (i.e., hydrostatic test) before coating and shipping the system. This standard requires that at least one coat of paint be applied to the tank and the dike. In addition, if the dike will not rest directly on soil, then appropriate corrosion protection measures should be taken (refer to STI R893 recommended practice).

Photo 12-1 shows an open-top diked AST.

Steel Tank Institute Standard F921-93

Steel Tank Institute F921-93, *Standard for Aboveground Tanks with Integral Secondary Containment,* First Edition, became effective in 1993. The scope of this national standard relates to the construction of steel aboveground integral secondary containment tanks with built-in monitoring capability. This secondary containment tank will be a Type I design, which entails a primary tank surrounded by a separate and distinct secondary steel shell that is an integral part of

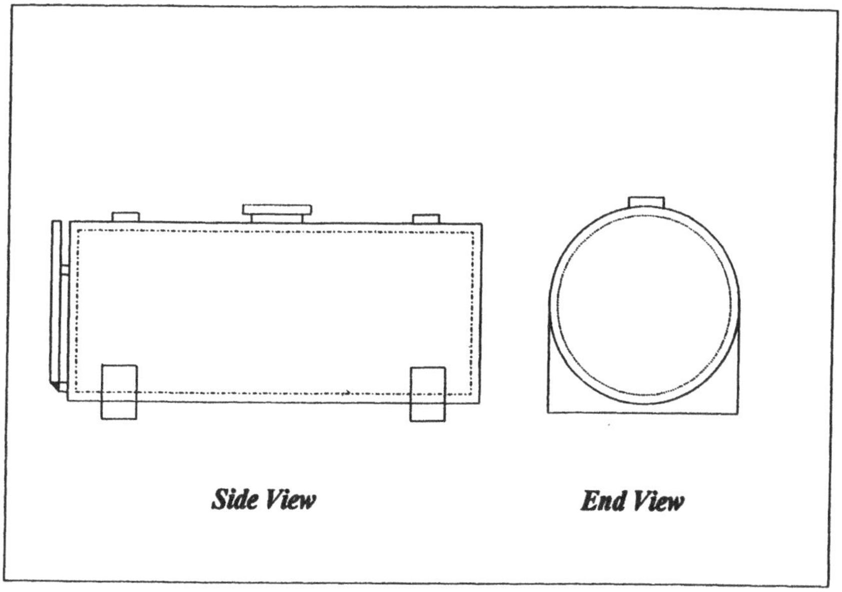

FIGURE 12-2.
Horizontal double wall AST (intimate wrap).

the primary tank allowing for containment of the contents of the primary tank during a product release. The F921 standard addresses both horizontal and vertical configurations.

Horizontal Double Wall ASTs

The construction of the secondary tank shall be distinct but in intimate contact with the primary tank, and the secondary tank shall envelope the entire 360 degrees of the primary tank, as shown in Figure 12-2. The primary tank has maximum sizing values based on nominal capacity, as highlighted in Table 12-2.

Table 12-2. Primary tank sizing allowances

Nominal capacity of primary tank (gallons)	Maximum diameter of primary tank* (inches)
up to 550	48
551–1,100	64
1,101–9,000	76
9,001–35,000	144

Source: STI F921-93 Standard.

*No primary tank can exceed 144 inches in diameter at any time.

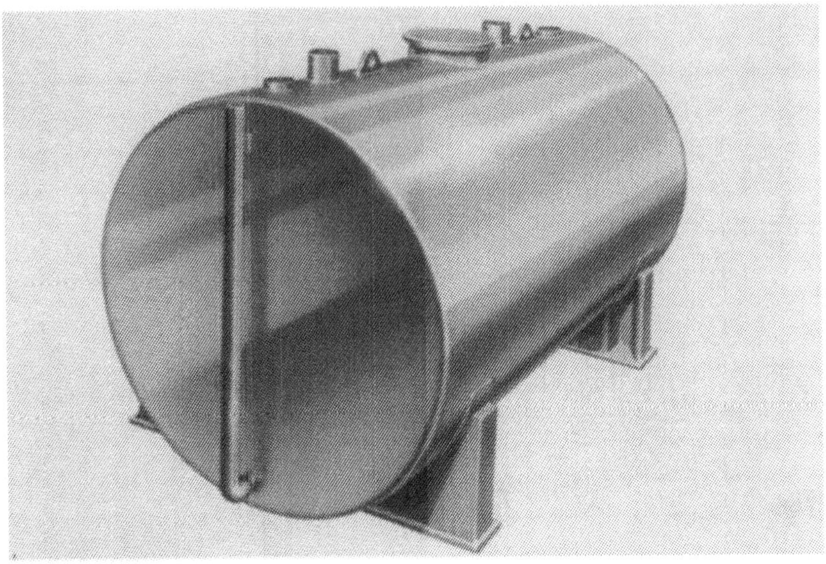

PHOTO 12-2.
STI-F921 horizontal double wall AST. (Source: Steel Tank Association)

Photo 12-2 shows a horizontal double wall AST.

Vertical Double Wall ASTs

The construction of the secondary tank shell and floor are to be distinct but in intimate contact with the primary tank, as shown in Figure 12-3. The roof connection (i.e., shell-to-roof joint) is to comply with the requirements of UL 142, including allowing a weak shell-to-roof joint as an acceptable means of emergency venting. Also, the thickness of the primary tank—including the shell, top, and bottom—shall comply with UL 142.

Photo 12-3 shows a vertical double wall AST.

General Requirements of Double Wall ASTs

The F921 standards recommends that manways be installed on tanks larger than 550 gallons, but requires manways on all tanks exceeding 5000 gallons. Monitoring of the interstitial space of the double wall AST is accomplished by adding an external monitoring pipe on one end of the tank attached through the secondary tank at a point two inches from the bottom of the tank (refer to Figure 12-2 for

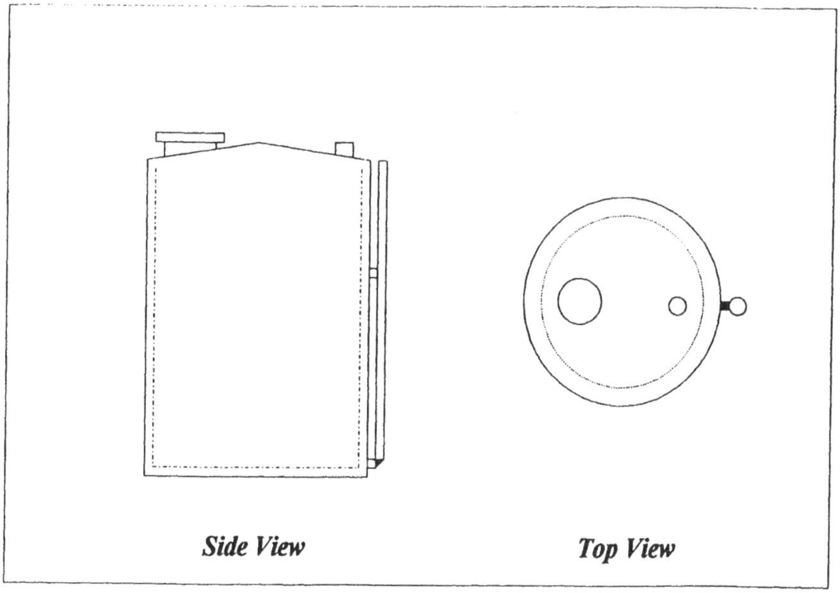

FIGURE 12-3.
Vertical double wall AST.

horizontal tanks and Figure 12-3 for vertical tanks). This standard lists four ways of monitoring the annular space:

1. Visual inspection.
2. Vacuum monitoring.
3. Mechanical float devices.
4. Electronic monitoring.

Tank penetrations, in general, should be per UL 142, and lifting lugs are to be installed on the top of the tank. All integral tank supports, such as steel saddles, are to be UL-listed. The F921 standard requires inspection and testing as well as venting and labeling of the tank system to be per UL 142 (Chapter 11).

Steel Tank Institute Standard F941-94

Steel Tank Institute F941-94, *Fireguard™ Thermally Insulated Aboveground Storage Tank Standard,* First Edition, became effective

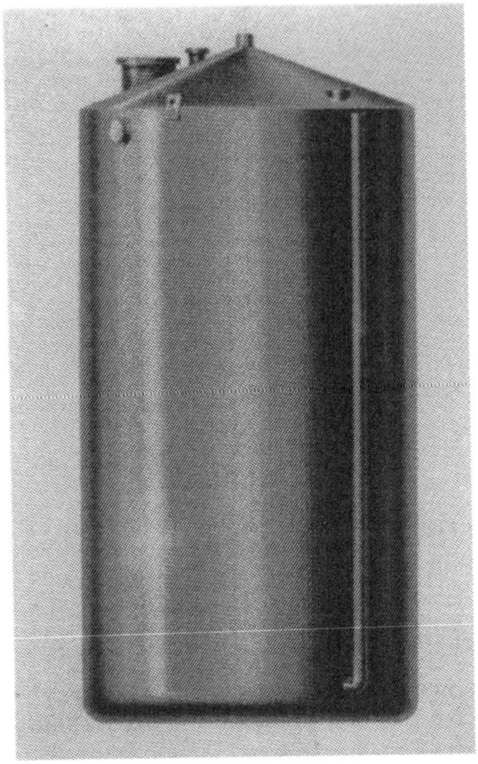

PHOTO 12-3.
STI-F921 vertical double wall AST. (Source: Steel Tank Association)

in 1994. The scope of this national standard relates to the increasing demand for fire-protected-tank technologies in the United States. This standard specifies applying thermal insulation to the interstitial space of a double wall aboveground storage tank to provide two-hour fire resistance. This tank system will bear the UL 2085 label for "an insulated secondary containment aboveground tank for flammable liquids." The F941 standard addresses both cylindrical and rectangular tank configurations.

Only manufacturers licensed by the Steel Tank Institute may manufacture a Fireguard™ AST. STI obtained the UL 2085 listing for its licensees, so every manufacturer must comply with the requirements of this standard or risk losing its license. Figure 12-4 illustrates the Fireguard™ design. (See also Photo 12-4.)

The F941 standard states that both the primary and the secondary tank shall be built to UL 142 specifications, allowing for a six-inch gap in the interstitial space. An internal monitoring tube for leak

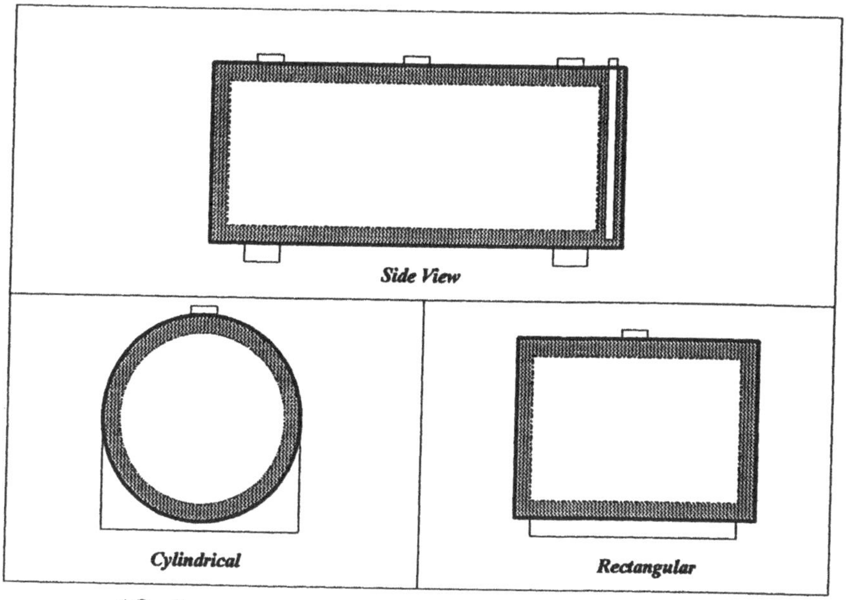

FIGURE 12-4.
Fireguard™ design.

detection purposes will be added against one head in the interstice. This standard has distinctive capacity requirements per Table 12-3 for both cylindrical and rectangular tank configurations.

Table 12-3. F941 capacity chart

Tank design type	Size range (gallons)
Cylindrical	250–50,000
Rectangular	500–2,000

Source: Developed from STI F941-94 Standard.

Every F941 AST over 5000 gallons shall include a manway to the primary tank. This would include every compartment exceeding a 5000-gallon capacity in a multi-compartment tank design.

Testing

As in other STI codes, the F941 standard requires air testing of the primary tank and interstitial space before shipment. Before application of the secondary tank, the primary tank shall be tested for leakage by applying 3.0 to 5.0 psig of air pressure while administer-

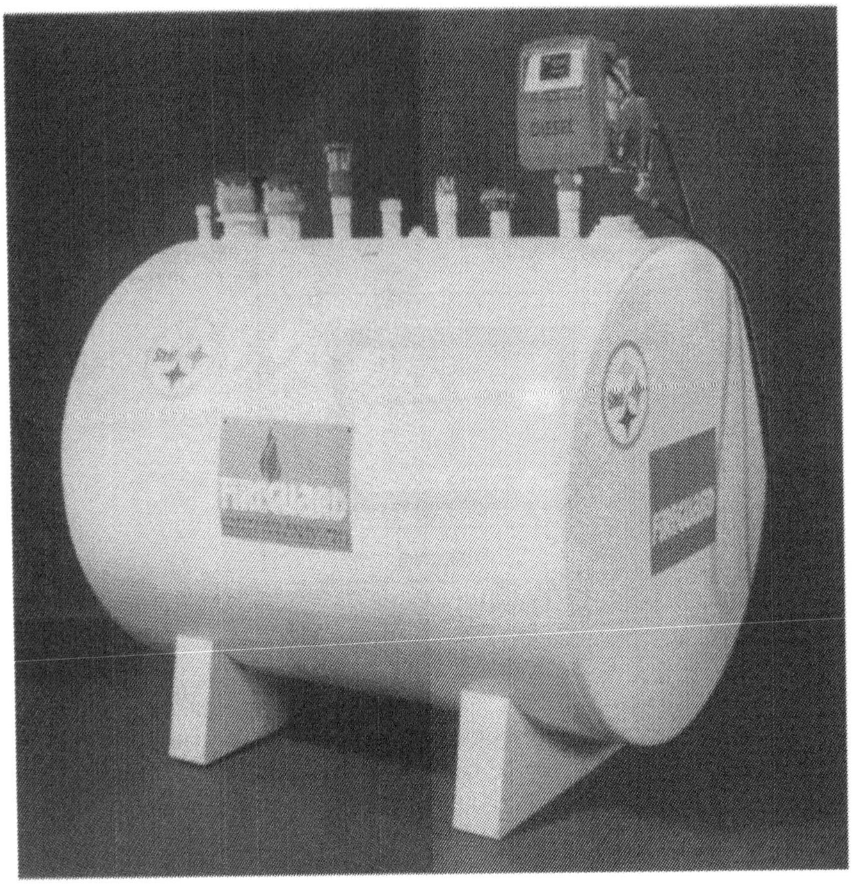

PHOTO 12-4.
STI-F921 Fireguard™ fire protected AST. (Source: Steel Tank Association)

ing a soap solution to the weld joints. After construction of the secondary tank, 3.0 to 5.0 psig of air pressure is reapplied to the primary tank, and then the air is bled off from the primary tank to the interstitial space of the two tanks until the interstice reaches 3.0 to 5.0 psig. Once again, a soap solution is applied to the weld joints of the secondary tank to check for leakage.

Insulation and Surface Coating

Upon successful leak test results, six inches of monolithic insulation material shall be applied to the interstitial space to provide for fire resistance of the tank assembly.

Coating of the exterior of the tank may be done before or after application of the insulation material to the interstice. The exterior of the tank shall be sandblasted and coated with 1.5 mils dry film thickness (DFT) of primer and then overcoated with 1.5 mils DFT of an enamel or a polyurethane type paint. Alternatively, a 3.0-mil DFT one-coat paint system is allowed.

Steel Tank Institute Recommended Practice R893-89

The Steel Tank Institute R893-89, *Recommended Practice for External Corrosion Protection of Shop Fabricated Aboveground Tank Floors,* became effective in 1989. A potential problem area for ASTs is corrosion on parts of the AST system that contact the soil. This recommended practice generally is for the bottom of vertical ASTs and the bottom of a steel dike. The method used to curtail corrosion in these areas is to utilize corrosion control techniques such as protective coatings, electrical isolation, and cathodic protection.

Protective Coatings

The exposed surface of the tank or the dike shall be blasted and coated on the full bottom and three inches up the shell or the side wall. The exposed surface is to be commercially blasted according to the Steel Structures Painting Council (SSPC) SP-6, Commercial Blast Requirements. The appropriate coating, per the manufacturer, is to be applied at 15 mils DFT minimum for AST systems that will be installed directly on soil. If the installation is to be on concrete or asphalt, a minimum 40-mil DFT coating must be applied.

Electrical Isolation

The purpose of electrical isolation is to completely isolate the AST system from its surroundings so that the system does not attract any stray electrical current that could lead to corrosion. Other structures exposed to the soil, such as piping and catwalks, would affect the AST system if not isolated. The recommended way to isolate piping is to use flange isolation kits.

Cathodic Protection

Cathodic protection of the exposed surface of the AST system is accomplished by installing sacrificial anodes (preferably magnesium anodes) varying in weight and size from 9.0 pounds to 48.0 pounds. Proper cathodic protection of vertical AST bottoms up to 164.0 square feet can be accomplished by using two magnesium anodes, with three magnesium anodes required for bottoms greater than 164.0 square feet. For dike bottoms, two to five magnesium anodes would be required, based on the square footage of the exposed surface area. Anodes should be buried in the soil at least 18 inches deep and spread evenly across the exposed surface area.

PART IV

AST Products

CHAPTER 13

New Field-Erected Aboveground Storage Tank Products

The ability to make a fully informed decision about the purchase of a new field-erected AST is of utmost importance. This chapter is intended to provide a basis for such decision making. For simplicity, the term "purchaser" is used to describe any party who is specifying an AST, who could be the owner or the operator of a facility or an engineer under contact to specify a tank for an owner or an operator. The chapter discusses three AST components: shell, bottom, and roof. The API 650 standard requirements are the focus of the discussion, with references to AWWA and API 620 made as appropriate for the subject matter.

The various standards include a separate section to assist the purchaser in specifying AST requirements to the manufacturer. API 650 has a four-page Appendix L, and AWWA has Section IV. General information should include size of the tank with respect to tank dimensions or capacity, design parameters, tank accessories or apertures, and other specific requirements. The authors have written a three-page tank data sheet containing pertinent information to assist the purchaser in specifying an AST. This data sheet is contained in Appendix G. The API has recently published API 2610 to assist management of an AST facility with site selection, installation, and system operations. This standard is discussed in Chapter 9 and is an excellent reference document.

When specifying the tank size, the purchaser must decide whether the size (dimensions or capacity) is nominal or working. Nominal size is the overall tank dimensions and does not take into account any unusable capacity that exists during tank operation. There is unusable capacity at the base of the shell due to an outlet nozzle elevation that prevents removal of all the product from the tank. There also is unusable capacity at the top of the shell due to the installation of an internal floating roof, an overflow nozzle, or any required freeboard that will not allow the product to reach the top of the shell. An

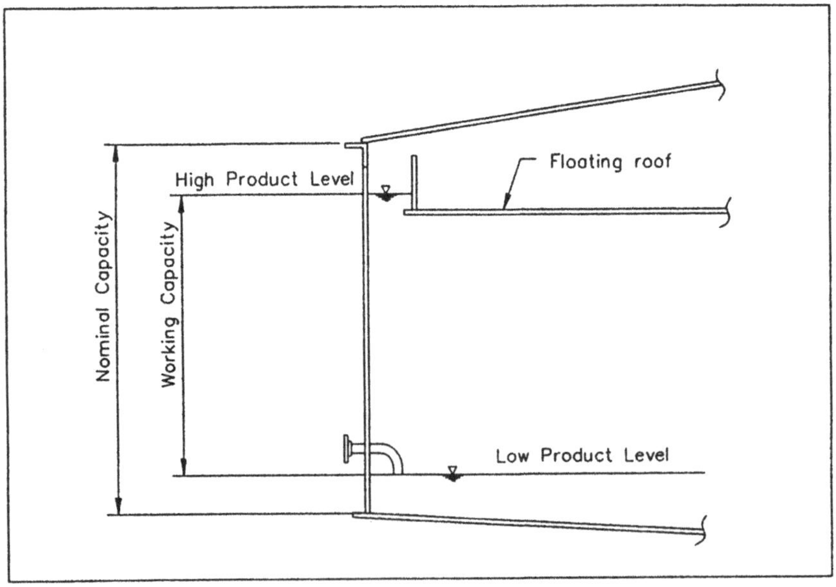

FIGURE 13-1.
Nominal and working tank capacity. The definition of working capacity varies with each AST owner/operator and must be specified to the manufacturer.

PHOTO 13-1.
Nearly-completed field-erected AST installation. (Source: Brown-Minneapolis Tank)

example of the difference between working and nominal capacity is given in Figure 13-1. If a working capacity is specified, the purchaser should include its definition, which varies with each owner or operator.

The second critical element in AST installation is knowledge of available design parameters such as specific gravity, vapor pressure, design temperature, and design pressure or vacuum, which are more important than the actual product being stored. Specifying these design parameters is critical, as they vary between products, and a purchaser may want the tank designed for some worst-case product parameters to allow flexibility in storing various products.

Finally, the AST purchaser should state applicable requirements of local and state codes to the manufacturer, as it would be impractical for a manufacturer to maintain updated regulations of every authority having jurisdiction. In addition, the manufacturer may be unaware of specific requirements placed by various agencies on an individual AST facility owner. The manufacturers of ASTs generally comply with state and federal laws, but the owner or the operator of a facility should be responsible for obtaining requirements of all other regulations and specifying them to the manufacturer.

Shell Design

In specifying that an AST be constructed according to the API 650 standard, three methods can be used to design shells of field-erected tanks. The proper method depends on tank size, allowing the manufacturer to utilize the most economical method. Appendix A is generally used to design tanks that are smaller and more economical than those of the main body of the standard when the determined bottom course shell thickness is less than the minimum plate thickness allowed per API 650. Table A-2 of API 650 shows the shell plate thickness for various combinations of tank diameter and height that could be used to approximate shell plate thickness. The main body of API 650 is used when the shell plate thickness is more economically determined by the one-foot or variable-point method. Either method will produce a thinner shell plate because of higher allowable stresss, but the benefit comes at the expense of additional testing such as radiograph examination. The largest increase in the number of radiographs between Appendix A and the main code occurs when the shell plate thickness is greater than 3/8 inches, which may be the best dividing line between the two different sections of the standard. The difference between the one-foot and the variable-point method is not so clear because both methods are based on the same allowable stresses and radiograph examination requirements. The difference between these methods is not evident until the tanks become large, generally greater than 100 feet in diameter. The variable method actually was developed for tanks with a large diameter, but can be used for an AST of any size with one limitation based on the tank diameter given in Section 3.6.4.1. Because economics plays the key role in the method of plate thickness determination, the decision should remain with the manufacturer unless the purchaser has a preference for one of the methods.

In designing an AST to the main body of API 650, the shell plate material chosen is critical to determining the most economical design. The most economical plate material is ASTM 36, which is classified as a medium-strength steel along with the other group I, II, and III materials given in Table 2-3 of API 650. Higher-strength materials, groups IV and higher, generally produce thinner shell plate. One detail that must be considered in material selection is to ensure adequate material toughness at the design metal temperature (DMT) specified by the purchaser. If a material does not meet the toughness requirements of API 650 Table 2-3, the material can be Charpy V-notch impact-tested to demonstrate toughness. Generally, only a fine-grain material, group IV or higher, has the capability of being Charpy V-notch impact-tested. These materials are generally more costly per pound, and the addition of impact testing only serves to increase the cost. However, the higher-strength material and the subsequently thinner plate may offset additional material costs.

In addition to material selection based on strength and cost, tank details affected by the material should be considered. For medium-strength material, any shell nozzle must be fabricated in either the shell plate or an insert-type reinforcing plate and stress-relieved when the shell plate is greater than one inch. The maximum thickness is decreased to one-half inch when the shell plate is a high-strength material. The stress-relief of shell nozzles will increase the cost of the tank and may increase fabrication time. In addition, the use of high-strength material requires a butt-welded annular ring under the shell for any shell thickness. Therefore, high-strength material will decrease the shell plate thickness, but the higher cost of the material, the requirements of impact testing, the annular ring, and possible stress-relieving shell nozzles must be factored in when one is making a final decision on the shell material type.

The majority of field-erected ASTs are constructed per API 650, but if the internal design pressure dictates a different code or if another standard is preferred, the AST can be constructed according to several other standards, as discussed in Chapters 9 and 10. To serve as a comparison, Table 13-1 outlines each of the methods of the various standards and the main parameters needed to determine the plate thickness using ASTM 36. The table lists the effective material allowable stress under tensile stresses and the MDMT of the plate allowed in order to meet the standard toughness requirements. If the pressure does not dictate the standard, Table 13-1 can be used to assist in selecting the most economical method, as the allowable stress is inversely proportional to cost.

Beyond the product properties required for shell design, the purchaser must specify seismic and wind loads that act on a tank. The application of national code seismic analysis such as that of the Uniform Building Code (UBC) is widely used to determine the seismic response of buildings, but is not recommended for flat-bottom tanks because sloshing of the product produces an increased moment at the tank base which these codes do not account for. The seismic design procedure used almost exclusively for ASTs is Appendix E of API

Table 13-1. Comparison of standards with the use of ASTM 36

Standard	Section	Joint efficiency[1]	Allowable stress[2] (psi)	MDMT
API 650	Main Body	1.0	23,200	10°F[3]
	Appendix A	0.70	14,700	-20°F
		0.85	17,850	
API 620	Main Body	0.85	13,600[4]	25°F[3]
		1.0	16,000[4]	
AWWA	Main Body	0.85	12,750	Unlimited
	Appendix C	1.0	19,330	20°F
ASME	Section VIII,	0.70	8,890	-20°F
	Division I	0.85	10,795	
		1.0	12,700	

[1] Joint efficiency is based on tension in the plate.

[2] The stated allowable stress is the allowable stress for ASTM 36 allowed by the standard and multiplied by the joint efficiency. The allowable stress also is based on design conditions at ambient temperatures, which may vary for elevated temperatures and other load cases such as hydrostatic testing.

[3] The standard or section alows the material to be used at varying MDMT, depending on the thickness. The MDMT listed is for 1/4-inch plate.

[4] The allowable stress is based on the plate under simple tension only.

650, and it is critical to specifying the seismic zone and associated site factors used in the analysis. If the purchaser requires that a tank withstand a certain wind speed, this too should be specified. If this information is not provided, API 650 allows the manufacturer to state the maximum wind speed that the tank will withstand before buckling. The wind and the seismic loads that act on an AST produce a moment at the tank base requiring the tank to be checked for overturning stability. A tank that will overturn under seismic loading is considered unstable and must be anchored unless certain characteristics of the tank are changed. For seismic loading, the resistance to overturning can be enhanced by increasing the bottom plate thickness under the shell and the subsequent effective product weight resisting the overturning moment. The API 650 standard limits the bottom plate thickness to an amount less than or equal to the bottom course shell thickness, and has restrictions on the contribution of the product weight. The seismic moment also produces compression stresses in the shell, which may cause the shell plate to buckle and thus require increased shell plate thickness. If the tank is still unstable with seismic loads imposed, it should be anchored or its dimensions changed to decrease the height-to-diameter ratio. The same changes, anchoring or altering the tank dimensions, should be made to the tank if it is unstable because of wind loads. The tank also should be checked to ensure that its shell will not buckle under wind loading, which could require a wind girder to stiffen the shell.

The installation of an AST most often requires the addition of secondary containment for spill confinement. As discussed in Chapter 3, on the Clean Water Act, secondary containment is required by the SPCC regulations, and the types of containment measures allowed include dikes, berms, and retaining walls. The EPA states that these structures must be impervious to oil for up to 72 hours and must be large enough to prevent the discharge of oil from reaching navigable waters while taking into account any collection of rainwater. The Oil Pollution Act of 1990 defines the size of secondary containment as 110% of the tank capacity. The discussion of secondary containment in the SPCC regulations does not actually size the diked area footprint. These sizing limitations are usually left up to the local fire codes, which most commonly refer to a national fire code such as that of the National Fire Protection Association (NFPA), and are discussed in Chapters 6 and 7.

The diked walls acting as secondary containment usually are constructed of earth, concrete, or a tank shell. Earthen dike walls usually are shorter than the other walls and located at a greater distance from the tank. Because an earthen dike will not be impervious to oil for up to 72 hours, the walls usually are lined. When the footprint of the diked area needs to be small, the dikes usually are constructed of concrete or steel. The use of a second shell for secondary containment can reduce the footprint of the diked area to just a few feet of the tank diameter. In this case the height of the steel dike wall will approach the tank height, and the tank roof may be extended to protect the diked area. The diked wall usually is constructed to the same standard as the tank so that the shell plate material and thickness are determined per API 650 and require the shell to be double butt welded. A steel dike wall is analogous to a double wall shop-fabricated AST except that field-erected tanks require a larger interstitial space to facilitate economic welding of the shell seams.

Bottom Design

The general design of a flat-bottom field-erected AST bottom constructed per API 650 or API 620 requires a single one-quarter-inch plate plus any specified corrosion allowance. API 650 requires that only the plate under the tank shell meet the toughness requirements of Section 2. This generally means that the plate material grade is the same as that of the upper shell courses because the plate is of nearly the same thickness. The standard is often incorrectly interpreted to mean that the bottom plate material should be the same as the bottom course shell plate.

Leaking AST bottoms have become an environmental concern because of contamination of groundwater and navigable waters, which has led to enactment of leak prevention and removal methods. The source of a leak can be attributed to two primary factors: the corrosion of plate and defects in the bottom plate or welds. In response to these problems, the use of double bottom tanks as part of a leak detection system has mushroomed during the past few years (see, e.g., Photo 13-2). Subsequently, API 650 added Appendix I, recommending types of release prevention barriers (RPBs) used for leak detection. Chapter 9, on API standards, discussed the two most widely used RPB designs in the AST industry, which are those constructed with a membrane liner and those made with a secondary steel bottom, as shown in Figure 9-3. Both of these systems are used primarily for their ease of installation and their relatively low cost to the owner or the operator. There are a number of advantages and disadvantages in using these two different systems, as outlined in Table 13-2. Chapter 14 examines the type of RPB best suited for existing tanks.

The EPA does not mandate that leak detection of the tank bottoms be implemented for ASTs storing oil products. However, it is the authors' impression that in the future leak detection of tank bottoms will be mandatory for all new and existing tanks, for a number of

Table 13-2. Advantages and disadvantages of two types of RPB for new tanks

System	Advantages	Disadvantages
Membrane Liner Figure 9-3(A)	• Liner will not corrode. • Medium between may facilitate cathodic protection.	• Cleanup costs higher if primary bottom leaks. • Slow movement of leaks to the perimeter.
Secondary Steel Bottom Figure 9-3(B)	• Leak detection is easier. • Cleanup costs are lower if primary bottom leaks.	• Accelerated local corrosion if space is unsealed. • Higher cost for interstitial space medium.

Source: API 650.

PHOTO 13-2.
Steel double bottom under construction for detection of leaks.

reasons. First, the CERCLA regulations for tanks storing hazardous substances require leak detection for new and existing ASTs. Second, some state or local environmental jurisdictions are requiring leak detection systems for new and existing ASTs, in line with a growing public environmental awareness. These actions may lead ultimately to greater federal or local government regulations, requiring leak detection for new tanks and mandating possible upgrades for existing tanks. Third, the Oil Pollution Act of 1990 requires that all oceangoing vessels be upgraded to have double hulls, and that the EPA study the requirement for AST tank bottom liners. Oceangoing vessels cannot be compared to ASTs, but in these current regulations the federal government has demonstrated the importance of preventing the discharge from any container from reaching waters or seeping into the ground, and it may ultimately require double bottoms or leak detection for ASTs. Finally, the cost of installing a secondary bottom for leak detection is minimal compared to the capital cost of the entire AST system or the potential liability of an oil or a hazardous substance discharge. These points underscore the importance of installing tank double bottoms now for leak detection—especially for new tanks—in anticipation of probable future requirements for secondary bottoms.

A number of factors need to be considered in selecting the leak detection system that best suits the AST operator's needs, including whether to have:

- Perimeter or entire bottom leak detection.
- Continuous or periodic leak detection.
- Electronic or manual leak detection.

A variety of systems are available to meet these requirements, most of them based on one or more of three human senses: sight, sound, and touch. The simplest method of leak detection is periodic visual inspection of the perimeter leak detection points for signs of a possible leak, and this can easily be fitted into the operating or maintenance plans of an AST facility. Another widely used system is acoustic emissions (AE), listening for leaks using highly sensitive sound sensors. Because the tanks must be isolated to remove background noises, this test is conducted only periodically. Two other types of leak detection systems can be operated continuously: the first uses electronic wiring installed under the primary bottom, which electronically senses a leak when the leaked product completes the circuit between two wires; and the second uses sensors to detect gases at the tank perimeter leak detection ports. These gases could be produced by the product being stored in the tanks or by gases introduced into the tank for the purpose of conducting the test. Before selecting a leak detection system, one should consult the federal, state, and local jurisdictions that require the leak detection system to determine their specific requirements.

Vacuum box testing of the tank bottom floor seams, which has long been a requirement of API 650, and adequate welding of the floor seams may have a preventive effect, minimizing the potential for leaks and the need for a leak detection system. One method of leak detection utilizes a vacuum testing procedure that doubles as a leak detection method. Here leak detection is achieved by pulling and maintaining a vacuum between the tank bottoms to ensure that no leaks exist in the bottoms following construction. If the vacuum deteriorates after a period of time, this change signifies that there is a leak into the interstitial space from either the primary or the secondary bottom. This type of system is described as a leak detection and prevention system because the vacuum applied between the bottoms is used as an actual testing procedure.

Because a smart, proactive approach to leak detection lies in prevention, the API has developed two standards for the implementation of leak prevention measures: API 651, *Cathodic Protection of Aboveground Storage Tanks,* and API 652, *Lining of Aboveground Petroleum Storage Tanks.* The guidelines contained in API 651, discussed in detail in Chapter 9, assist the owner in determining whether corrosion of the tank bottom underside is a problem and the best type of cathodic protection to install. Interstitial space cathodic protection can be implemented in a double bottom design to prevent the corrosion of the primary bottom as well. For this type of application, the best design of the double bottom uses sand between the bottoms to allow proper operation of the cathodic protection system. The capital cost of installing cathodic protection is small in comparison to the cost of the total AST system and the potential liability involved following a leak. Therefore, the authors recommend that cathodic protection be utilized for new tanks if the foundation design requires the application to prevent corrosion.

Corrosion of the tank bottom plates also can occur inside the tank, as a result of the product's being stored, or at the interface of the petroleum and water that has collected on the tank bottom. For this reason, the owner or the operator of an AST may want to consider the addition of a liner to the tank bottom and a small portion of the shell to prevent corrosion of the shell plate. Chapter 9 discussed the use of API 652 to assist in the selection and the application of a tank liner for this purpose. Two different types of liners can be used if there exists a potential for internal corrosion based on past experience.

Roof Design

A tank only requires that a wind girder be installed to keep the tank shell round and prevent it from buckling under wind loads. Therefore, a tank roof is not an essential variable, but one may be used for many reasons, including:

- Keeping rainwater, snow, and other foreign elements from contaminating the product.
- Preventing loss of product due to evaporation.
- Reducing the emissions of air pollutants.
- Maintaining pressure as required for the storage of certain products.

Thus, an owner or an operator typically is motivated to install either a fixed or a floating roof. Additionally, the EPA has tightened regulations to eliminate leaks into the ground from waste treatment lagoons by requiring secondary containment, which often leads to the use of open-top, fixed or floating roof ASTs. Applications such as storm water holding tanks or final clarifiers of a waste treatment facility may not require the use of a tank roof.

If a tank roof is required, there are many roof options or types on the market. In selecting a roof, there are a number of factors that one must consider:

- The most economical of the different roof types for a given tank size.
- The best design to facilitate the application of a paint system.
- The amount of any internal pressure.
- The corrosion allowance required on the roof system.
- One's preference with regard to internal maintenance and cleaning.

These factors should be used by the owner to eliminate roof design options, with final roof design selection based on economics and left up to the manufacturer following a cost analysis. If the purchaser has a preference for a specific roof design, this should be clearly stated. The general rule for selecting the most cost-effective roof is to choose a self-supported cone for tanks less than about 25 feet in diameter and a supported cone roof for those greater than 25 feet. When internal structural members cannot be used for reasons such as the use of an internal paint system, the tank should be designed by utilizing an umbrella or a dome roof. Umbrella roofs are single-curvature, a design that is economical and can be used on tanks with relatively low pressure. Dome roofs are double-curvature, a more costly option, and should be reserved for tanks under significant pressure. A new addition to the API standard is Appendix G, which outlines the design of aluminum geodesic dome roofs. (See Photo 13-3 for an example.) From a maintenance standpoint, aluminum roofs may be the best choice because aluminum is generally corrosion-resistant and often does not require painting.

Internal pressure in an AST usually is related to the product being stored. A principal source is pressure exerted above the product to minimize product evaporation due to product vapor pressure. Also, the tank may be designed to hold the product vapors, thus eliminat-

PHOTO 13-3.
Aluminum geodesic dome roof field-erected AST.

ing breathing losses that normally would escape into the atmosphere. Further, the volume above the product sometimes is purged with an inert gas such as nitrogen, to reduce the risk of explosion of product gases. The inert gas is kept under a slight pressure, in an operation commonly referred to as a nitrogen blanket. The pressure required to keep the product in the tank liquid or to reduce the amount of evaporation generally governs the tank design standard. Tanks with a low pressure, such as those with nitrogen blankets, generally are designed according to the API 650 standard. When the product is very volatile in the liquid state, the use of API 620 or ASME pressure vessels is recommended to keep the product in liquid form and eliminate evaporation losses.

The amount of internal pressure can influence the type of roof design. Generally, the internal pressure will increase the area at the shell–roof intersection and possibly increase the roof slope. For example, tanks over 25 feet in diameter generally remain supported cone roof tanks for small pressures, but the roof slope may change from the standard of 3/4 inch in 12 inches to the steepest roof slope allowed per API 650, 2 inches in 12 inches. This will increase the angle at the shell–roof intersection and subsequently reduce the required area at the top angle. API 650 Appendix F allows for an internal pressure design of up to 2.5 psig. However, when the internal pressure acting on a self-supported roof exceeds the nominal weight of the tank, API 650 requires that the roof and the shell–roof intersection be designed per the rules of API 620. This standard requires the roof to have double curvature, effectively disallowing the use of an umbrella roof. Given all these factors and design criteria, the selection of a roof system can be complicated, and it must be weighed carefully by both the purchaser and the manufacturer.

The venting requirements of any tank, particularly a pressure tank, are critical to proper operation. For pressure tanks, the system for the most part is closed to the atmosphere but must be vented to ensure that the tank does not overpressurize and cause structural damage. Overpressure of tanks causes buckling of the top angle region or, worse, separation of the shell-to-bottom joint, resulting in spillage of the tank contents. To adequately specify and design a pressurized tank, one must fully understand the operation of vents. All pressure and vacuum relief vents have a set pressure at which the vent will open. To obtain adequate or full venting capacity, the pressure inside the tank must increase above the set pressure by a significant amount. Therefore, it is critical to ensure that the tank is not designed for the set pressure but rather for the pressure required to obtain adequate venting flow rates.

Beyond the pressure buildup due to normal operating conditions, the AST may exhibit a pressure due to emergency pressure relief if the AST does not have a frangible joint per API 650. As discussed in Chapter 9, when the area of the top angle region is greater than that allowed for a frangible joint, the AST is not frangible, and emergency venting should be installed. Because most tanks are designed with the minimum angle required by API 650 or the required area per Appendix F due to internal pressure, it is typically not economical to alter the tank to meet the frangible joint requirements. Typically, the tank diameter should be in the 50- to 75-foot diameter range or greater before the tank will have a frangible joint.

Vacuum loading or external pressure can cause significant damage to ASTs. The results of a vacuum load on a tank are analogous to the physics experiment of sealing an aluminum can of hot water and watching the side walls buckle as the water cools. The same mode of failure will occur when the tank is closed to the atmosphere, and the contents cool down or product is withdrawn. A large shell may be able to withstand a large internal pressure, but if it is not properly designed for external pressure, the allowable pressure is very small. API 650 (Section 3.2.4) states that a tank will withstand an internal pressure of only one ounce per square inch, which is equivalent to about 2 inches of water or 0.15 inch of mercury, a magnitude of change in the atmospheric pressure that can be expected overnight. Therefore, it is critical that a tank be properly vented even when it is not in operation.

To ensure proper venting, API publishes standard API 2000, *Venting Atmospheric and Low-Pressure Storage Tanks,* to determine the required venting pressure and vacuum relief flow rates. The standard determines the required flow rate based on inbreathing and outbreathing due to product movement into or out of the tank and a change of state, such as from gas to liquid. If the tank requires emergency venting, the standard also specifies the required flow rate. For an atmospheric storage tank, normal venting usually is provided by a gooseneck vent, which is simply a roof nozzle and two 90-degree elbows to prevent moisture from entering the tank. When the tank is designed to hold pressure, the normal vent is a mechanical vent, constructed of weights or springs, that opens at a set pressure. Typically, the emergency venting, in addition to the normal venting required for these tanks, is often a mechanical vent attached to the roof manhole. It is important to note that a tank constructed to standard API 650 does not automatically require the manufacturer or any party to design the venting; so it is imperative that the purchaser clearly specify if the manufacturer or the process engineer is responsible for the venting design, in addition to noting inflow and outflow rates.

Floating Roof Design

Floating roofs for ASTs originally were intended to prevent loss of product due to evaporation. With an increase in air pollution, federal and state laws have progressively changed over the years, requiring controls on storage tanks to reduce the emission of certain pollutants and adding another purpose to the use of floating roofs. As discussed in Chapter 4, the Clean Air Act requires new ASTs that are greater than 10,500 gallons in size and storing product of a given vapor pressure to have some type of control to prevent product emission. The type of control is either an internal floating roof (IFR), a vapor

PHOTO 13-4.
Internal floating roof field-erected ASTs under construction.

recovery system for fixed roof tanks, or an external floating roof (EFR). The decision to use either type of floating roof or a vapor recovery system is based on the product stored and the feasibility of vapor recovery. This section examines the two different types of floating roofs and will assist the purchaser in making an informed decision on the type of floating roof to install. (Photo 13-4 shows IFR ASTs under construction.)

The primary difference between an internal floating roof and an external floating roof is that an external floating roof does not need a fixed roof; therefore, the two types differ with respect to initial capital cost, protection from the natural elements, and maintenance. The cost differences between internal and external floating roofs are small, but the cost of the AST is much lower for an external floating roof tank compared to an internal floating roof tank because of the additional fixed roof. However, the fixed roof does protect the internal floating roof from the elements (e.g., wind, which can cause damage to an external floating roof positioned near the top of the tank, and lightning, which can cause a seal fire). Moreover, a fixed roof drains rainwater naturally, but an external floating roof relies on the proper operation of a roof drain system. The maintenance of fixed roofs has become a critical consideration in choosing a floating roof system. With the safety requirements associated with entering a tank, an external floating roof offers better visual inspection of the floating roof and safer access than a fixed roof tank containing an internal floating roof.

A facility may be under strict emission restraints and may be required to examine emissions from both internal and external floating roof tanks. As discussed in Chapter 9, API publishes a number of standards dealing with emissions from ASTs. There are standards for the emissions from internal floating roofs (API 2519, *Evaporation Losses from Internal Floating-Roof Tanks*) and external floating roofs (API 2517, *Evaporation Losses from External Floating-Roof Tanks*). Both standards separate evaporation losses into standing losses and withdrawal losses. Standing losses are associated with the floating roof construction, type of accessories, and seal type, and make up the majority of losses for either type of floating roof. Withdrawal losses are a function of tank operation, in particular the annual product throughput, and are usually small in comparison to standing losses. Therefore, a comparison of internal and external floating roofs usually is based solely on standing losses. Both standards list evaporation losses from a number of different accessories, independent of the stored product, and give an example of evaporation loss calculations for a 100-foot-diameter tank, which is the basis for Table 13-3. The units of evaporation loss in Table 13-3 are pound-moles per year, which are used for comparison of the different floating roofs but do not represent actual losses. To determine the actual losses for a given tank, these values must be multiplied by factors dependent on the product being stored: vapor pressure, molecular weight, and a product factor. The comparison was made by using comparable and standard equipment for both types of floating roofs and shows that the losses are less for an internal floating roof, even with a secondary seal installed in the external floating roof.

After determination of whether an AST will be outfitted with an internal or an external floating roof, two different types of external floating roofs and six types of internal floating roofs must be considered. An external floating roof is typically an annular pontoon or a double deck, with the advantages of each type differentiated by cost analysis and rain drainage. Although the cost of a double deck can be much higher than that of an annular pontoon because of the addition of a second deck, the double deck roof will offer better drainage of rainwater because it has top plates that are sloped to the center rather than a single plate resting on top of the product. A

Table 13-3. Vapor losses from external and internal floating roofs

	EFR evaporation losses[1] (pound-moles per year)	IFR evaporation losses (pound-moles per year)
Accessories[2]	68	138
Roof legs	74	253
Guide pole[3]	640[4]	57[5]
Seal	200[6]	300[7]
Total	992	760

Derived from: API 2517 and 2519.

[1] Losses from EFR are based on 10 mph wind velocity. [2] Required accessories that include: manhole, vacuum breaker, float well, column seal, rim vents. [3] Ungasketed gauge well. [4] Unslotted gauge pipe. [5] Slotted gauge pipe. [6] Mechanical primary shoe seal with rim mounted secondary seal. [7] Liquid mounted primary seal.

double deck floating roof also has a greater number of pontoons to prevent the floating roof from sinking.

Because of the large number and greater disparity of internal floating roofs, the basis of selection is broader than that for external floating roofs. To facilitate comparison of the different types, the authors have chosen four comparative factors. First is the typical material of construction, usually carbon steel or aluminum. Aluminum is beneficial in some applications because of its protection against corrosion with no need for additional coating. Second is the number of pontoons, as pontoons can puncture or leak during the lifetime of the floating roof, and a single pan or pontoon would sink. This can happen if the deck or rims are punctured or if product gets on top of the deck and outweighs the roof. Therefore, a multiple pontoon floating roof offers the greatest stability even when a few pontoons become flooded. Third is covering of the pontoons, as pontoons that are covered or closed offer the benefits of increased protection against pontoon flooding and lower maintenance. Fourth is contact with the products, as a floating roof in complete contact with the product surface will allow a minimum amount of product vapors to be collected under the roof, helping to eliminate a hazard to maintenance personnel and the source of a possible explosion. Table 13-4 outlines the different types of internal floating roofs and basic features associated with each type. Two other factors may influence the decision on what type of floating roof to select: emissions and economics. If the selection is based on emissions or

Table 13-4. Specific features of internal floating roofs

Type[1]	Typical material of construction	Multiple pontoons	Closed pontoons[2]	Full surface contact[3]
Pan	Carbon steel			
Compartmental	Carbon steel	√		
Annular pontoon	Carbon steel	√	√	√
Double deck	Carbon steel	√	√	√
Deck-on-floats	Aluminum	√	√	
Sandwich panel	Aluminum			

[1] The types of floating roof are graphically described in Figure 9-2. [2] Pontoons are closed and will not allow product or other debris inside. [3] The deck is in full contact with the surface of the product.

vapor losses, API 2519 states factors to consider if the deck is constructed of welded or bolted seams. There are no losses from welded seams, but losses from bolted seams utilized in aluminum floating roofs are equal to 0.068 pound-mole per foot-squared-year for an unknown width of plates used. For a 100-foot-diameter tank, these losses would equal 680 pound-moles per year or nearly 50% of the total losses when combined with the losses given in Table 13-3 for an internal floating roof. If pure economics are used to select a floating roof type, the manufacturer should be asked to conduct a capital cost–benefit analysis of each type, and the owner should factor in the annual operating costs over the floating roof's lifetime to make a complete and informed decision.

In addition to the choice of a floating roof, the appropriate perimeter seal must be selected. Three primary types of seals are utilized: a bag seal, which is a foam- or liquid-filled fabric; a mechanical shoe; or a wiper type seal. A bag seal has long been the primary seal used, as either a primary or a secondary seal. The seal is relatively inexpensive but may not last as long as others and can be difficult to maintain. A primary liquid-mounted mechanical shoe seal is more expensive than a bag seal and varies in style and longevity between manufacturers, but generally will last longer. A wiper seal can be used for either a primary or a secondary seal and is generally some type of extrusion seal, such as rubber, mounted either on the floating roof rim or on a plate attached to the rim plate, which also acts as a weather guard for the primary seal.

There are two primary criteria for the selection of a seal system for a floating roof: economics and emissions. Because the seals wear against the shell as the tank is emptied and filled, the equivalent annual cost over the lifetime of the seal must be considered. Therefore, the selection of the seal system must be based on a cost analysis of not only the equivalent annual capital cost but also the maintenance and disposal costs when the seal or parts are removed. These costs may be high if the seal, in particular the foam core of a bag seal, is saturated with product. The type of seal utilized on a floating roof also may be dictated by emission controls or levels of the particular AST facility. The emission controls required of AST facilities may be dictated by the different government agencies with jurisdiction over the facilities. Chapter 4 explained that the Kb controls specify which type of seals must be placed on certain ASTs, and Chapter 8 examined further control restrictions required by state and local government. A facility also may be under parameters established by governing bodies to regulate the amount of vapor that can be emitted into the atmosphere. Table 13-5 shows the amount of evaporation losses that can be expected from the various types of seals. These are taken from API 2517 and API 2519 for a 100-foot-diameter tank and generally are losses independent of the product being stored.

Table 13-5. Evaporation loses from seals

Floating roof type	Seal arrangement[1]	Evaporation losses[2] (pound-moles per year)
Internal floating roof[3]	Liquid-mounted primary	300
	Liquid-mounted primary with secondary	160
	Vapor-mounted primary with secondary	250
External floating roof[4]	Mechanical shoe primary with secondary	200
	Liquid-mounted primary with secondary	176

Derived from: API 2517 and 2519.

[1] Secondary seals are all rim-mounted. [2] Based on an average fit of the seal on a 100-foot-diameter tank. [3] Evaporation losses based on API 2519. [4] Evaporation losses based on API 2517, at a wind velocity of 10 mph.

A final consideration in the purchase of an AST concerns proper overflow protection, especially in an AST outfitted with a floating roof. Proper overflow guarantees that the floating roof will not incur any damage as a result of overfilling of the tank. For internal floating roof tanks, API 650 Appendix H specifies that the tank have overflow slots or be equipped with appropriate alarm devices. Overflow slots are openings in the shell designed to discharge product at the pump-in rate. This type of discharge will constitute a spill according to the SPCC regulations discussed in Chapter 3. Therefore, ASTs increasingly are being outfitted with overflow piping and/or adequate alarms to prevent damage to the floating roof. The API publishes a standard, Recommended Practice API 2350, *Overfill Protection for Petroleum Storage Tanks*, for the design of overflow alarm systems. This should be used to design overflows for floating roofs with or without overflow slots to prevent the discharge of product in violation of the SPCC regulations. The actual overflow elevation depends on the type of seals used and the floating roof manufacturer.

CHAPTER 14

Existing Field-Erected Aboveground Storage Tank Products

As the AST market is becoming more strictly regulated by various governing agencies, the structural condition of existing ASTs is becoming an increasingly important issue for owners and operators. As discussed in Chapter 13, current regulations require secondary containment of ASTs, and future expanded regulations may mandate the new and existing ASTs be outfitted with some type of leak detection system. Even if such regulations are not developed, the OPA of 1992 and other regulations set high liability limits on the discharge of oil or other hazardous material. The costs associated with maintenance and modifications of ASTs may be significantly less than the potential costs of removing an oil discharge.

In addition to federal and state regulations requiring modifications, the inspection or a planned change of service of a tank may necessitate repairs or modification. To ensure that these repairs are made according to good engineering practice, API 653, *Tank Inspec-*

tion Repair, Alteration, and Reconstruction, is available to assist an owner or a manufacturer in the repair work. This standard outlines the requirements of a proper inspection, including specific inspection details, evaluation of the inspection report to determine any required changes or modifications, and associated procedures to follow in conducting repairs. This chapter focuses on examining the condition of existing ASTs and on associated repairs or alterations required for continued service or change of service. The chapter is divided into four sections: inspection, evaluation, repairs and modifications, and reconstruction.

Inspection

The single most important component of AST operations is suitable inspection, at appropriate intervals, to assure proper tank operation. Two types of inspection are important to AST operations: routine and scheduled. A routine inspection should be conducted by the AST operator daily or at the start of a shift. The inspection should be of the AST exterior to ensure that the facility is operating properly and that there are no obvious problems with the tank system. The second type is a scheduled inspection by a qualified inspector at predefined intervals. This inspection is conducted on the AST exterior and interior to ensure that there are no potential problems and to determine whether any physical or operational changes are necessary. The main purpose of this inspection is to determine the amount of corrosion that has occurred since construction of the tank or the last inspection. To assist the qualified inspector and the owner in setting up an inspection program, API 653 Appendix C contains a checklist for both in-service and out-of-service inspections.

The key to a scheduled inspection is the time interval between inspections of both interior and exterior surfaces. The monetary cost of an inspection, which may include the loss of revenue and the cost of cleaning the tank for safe entry, can be high. Therefore, the interval should be as long as possible to avoid excessive inspection costs over the lifetime of the tank, but frequent enough for the operator to make modifications to ensure that the tank will not develop problems. API 653 (Section 4) outlines the required intervals of inspection, which are determined by a variety of factors including:

- The nature of the product stored.
- The results of visual inspection and previous inspections.
- Corrosion considerations, including allowances, rates, and corrosion prevention systems.
- The methods and materials of construction and repair.
- The location of the tank and the associated risk to the environment.
- Leak detection and secondary containment systems.
- State and local regulations.

For external inspections, the API 653 standard suggests monthly routine visual inspections of the tank exterior and visual inspections by a qualified inspector at least every five years or at every quarter of the corrosion life of the shell. Further, the inspection of an AST should include periodic in-service ultrasonic thickness measurements of the shell to determine the rate of general uniform corrosion. The extent of ultrasonic measurements shall be determined by the owner/operator and at a minimum should include frequent readings of the bottom courses and along four vertical lines on the remaining courses. According to API 653, this should be done five years after the commission of the tank, or when the corrosion rate of the tank is unknown. When the corrosion rate is known, the interval should be one-half the corrosion life of the shell but no longer than 15 years. A summary of thorough inspection intervals is given in Table 14-1 for both external and internal surfaces.

Table 14-1. Minimum qualified inspection intervals per API 653

Surface	Inspection type	Unknown corrosion rate	Known corrosion rate
Internal	Plate thickness	10 years	20 years
External	Visual	5 years	5 years or one-quarter of corrosion life
	Plate thickness	5 years	One-half of corrosion life or 15 years

Derived from: API 653.

An external inspection gives a good indication of shell corrosion, whereas an internal inspection is designed to verify the shell corrosion status and determine the bottom corrosion. API 653 does not specifically set intervals for internal inspections, but states that the interval should be based on the expected tank corrosion, particularly the rate of bottom corrosion. For this rate, subsequent intervals of internal inspections are developed, based on the minimum bottom plate thickness given in Table 4-1 of API 653 and with the minimum internal inspection interval not less than the values listed.

An important part of the inspection process is to maintain adequate original tank construction records and past inspection reports. Drawings and calculations from the original construction are required to determine the as-built shell thickness and design parameters. This information is used to determine the corrosion rate of the tank components and to evaluate the tank for continued use or change of service. Past inspection reports and the corrosion rates developed are crucial for operators, to further define and attempt to increase the inspection interval; so it is essential to maintain all records for the entire lifetime of the tank.

Evaluation

The information and measurements taken during an inspection must be evaluated to determine suitability for continued use or change of service as discussed in Section 2, *Suitability for Service,* of API 653. The primary emphasis of a tank evaluation is to observe the shell and bottom plate thicknesses for general and localized corrosion. General corrosion is the depletion of plate on a uniform basis, whereas localized corrosion is concentrated in a smaller area. Pitting of the shell is corrosion in very small areas, but, overall, it does not pose a threat to the structural integrity of the tank unless the pitted areas are in close proximity to each other. As stated in Section 2.3.2.2 of API 653, pitted areas may be ignored if the pit depth does not exceed one-half the minimum shell plate thickness, or the sum of the vertical dimension of close pitted areas does not exceed two inches in eight inches.

Section 2.3.2 of API 653 outlines a procedure used to determine the actual plate thickness of corroded areas based on a vertical length L. It is over this length that API 653 assumes that local hoop stress risers, resulting in plate discontinuity such as local corrosion, are averaged out, and this length is a function of the minimum plate thickness in a corroded area (t_2). A number of vertical plates are mapped out over the tank shell corroded area, with at least five thickness measurements recorded over the length L and the minimum average of these five readings over these vertical planes defined as t_1.

Once the actual plate thicknesses (t_1 and t_2) are determined, these values should be compared to the minimum thickness (t_{min}) as defined by Section 2.3.3 of API 653 and reproduced in equation (14-1). The value of t_{min} should be greater than the minimum average thickness (t_1), and the minimum thickness (t_2) should not be less than 60% of t_{min}. The minimum plate thickness shall also be checked for all imposed loadings per API 650.

$$t_{min} = \frac{2.6D(H-1)G}{SE} \tag{14-1}$$

where:

D = tank diameter

H = distance from the bottom of the length L for the most severe corroded area to the maximum design liquid level

G = specific gravity, to insure that this value accounts for any possible hydrotest

S = allowable stress, which should be less than $0.80Y$ or $0.426T$ for the bottom two courses and $0.88Y$ or $0.472T$ for the upper courses

Y = minimum yield strength or 30,000 psi if the shell plate is unknown.

T = minimum tensile strength, which should be less than 80,000 psi, or equal to 55,000 psi if the shell plate is unknown

E = original tank shell joint efficiency, or 0.70 if unknown, or 1.0 if one is evaluating minimum plate thickness away from shell seams

Equation (14-1) often is written to solve for the maximum design liquid level for a given minimum shell thickness, thus limiting the amount of product the tank stores. The product being stored also may be changed to a lower specific gravity to take advantage of the full tank capacity.

The shell plate must be evaluated to ensure adequate toughness, especially if the tank will undergo a change of service. The most dramatic change of service inolves a heated tank being converted to an ambient storage tank, making it critical to evaluate brittle fracture failure because of the lower temperatures the tank will experience. Section 3 of API 653, *Brittle Fracture Considerations,* defines the evaluation requirements for shell toughness outlined in the form of a decision tree (reproduced in Figure 9-4). The fundamental principle behind brittle fracture toughness is that the tank material must meet the toughness requirements of the latest edition of API 650, demonstrate toughness by undergoing a hydrotest, or be exempt. An exemption occurs when features of a tank exist that limit the exposure to brittle fracture failure, such as low stress in the plate (less than 7000 psi), a relatively thin shell thickness (less than or equal to 0.5 inch), or operation at elevated temperatures (greater than 60°F). For other situations, the API 653 (Figure 3-2) standard may exempt a tank of a given thickness and shell metal temperature from demonstrating toughness.

The minimum roof and bottom thicknesses allowed per API 653 are covered in Sections 2.2 and 2.4, respectively. When the average corroded roof plate is less than 0.09 inch per 100 square feet, API 653 recommends that the plate be repaired or replaced. The minimum bottom plate thickness depends upon whether the bottom is designed without tank bottom leak detection and containment or is lined in accordance with API 652. For tank bottoms without protection, the minimum plate thickness is 0.10 inch, whereas for protected tank bottoms (i.e., a leak detection system or a tank bottom liner), the minimum bottom plate thickness is 0.05 inch.

The bottom plate thickness can be determined by two methods according to API 653: the deterministic method described in Section 2.4.7.1 and the probabilistic method described in Section 2.4.7.2. The deterministic method is a complex set of equations based on thickness readings, usually measured by a complete scan of the floor. A complete floor scan will record the thicknesses of over 90% of the bottom and is usually completed by one of two methods: an ultrasonic floor scan or a magnetic flux leakage floor scan that measures the loss of the magnetic field in a plate due to plate thinning. The probabilistic method is a statistical analysis based on random single thickness readings of the bottom using an ultrasonic meter. Both methods will accurately determine the minimum average thickness, but the two methods differ in cost and accuracy in mapping out areas where the thickness is less than or near the minimum. The deterministic method is more costly, but locates all areas of any given thickness. In contrast, the probabilistic method is less expensive, but the analysis will determine the thickness at only a few locations and not outline the exact area of a given thickness. The danger is possibly missing areas under the minimum thickness or existing holes. Thus, the probabilistic method may be used first to determine the average thickness of the plate. If the thickness is near the minimum, or if it is known that there are areas where the plate is less than the minimum, the deterministic method should be used to pinpoint these areas and determine the extent of the corrosion. This type of information is required for one to perform a cost analysis and determine if a bottom must be replaced or repaired.

Repair and Modifications

Based on the inspection and the evaluation, the tank bottom and the shell may require repair or modification due to depletion of the plate as a result of corrosion or bottom settlement. In addition, government agencies may mandate leak detection or bottom secondary containment for tank bottoms. This section examines the possible types of repairs made to a tank and reviews the options available for obtaining the most economical solutions. The decision to repair or to replace usually is dependent upon the amount of corrosion and the associated costs of the options. A tank manufacturer experienced in tank repair can assist one in the cost comparison. (Photo 14-1 shows a field-erected AST undergoing repair.)

The primary reason for tank repairs or modifications is depletion of the bottom plate to less than the minimum allowed per API 653, with the type of repair depending on the extent of corrosion. As discussed above, if the average plate thickness is less than the minimum, the area of the bottom that is less than or near the minimum thickness can be determined by a complete scan of the bottom. From this information, a cost analysis of repairing the bottom should be conducted, comparing the repair cost to the cost of replacing the entire tank bottom.

If the tank bottom needs replacement, a number of different methods can be used. The two most common methods are slotting and jacking of the shell for the installation of the new bottom, pictorially described in Figure 14-1. The most common method, and the one recommended by API 653, is slotting of the shell about three to six inches above the old bottom and installation of a new bottom through the slot, extending it beyond the shell as required by API 650. The method of slotting, as described in Section 7.9.2 of API 653, recommends that a suitable noncorrosive material such as sand, gravel, or concrete be placed between the bottoms. Consideration should be given to installing cathodic protection between the two

PHOTO 14-1.
Open-top floating roof field-erected AST under repair and conversion to an internal floating roof AST.

bottoms, especially if the existing bottom is not removed. API 652 states that whether or not the secondary bottom is replaced, a double bottom design has a significant impact on the cathodic protection system. The second method of replacing the bottom amounts to removing the existing corner weld between the bottom and the shell or cutting the shell above the corner weld, jacking the shell, removing a small portion of the existing bottom under the shell, installing a new bottom on top of the existing one, and welding it to the shell. It is important to note that installation of a secondary bottom via any method may require modification of the shell nozzles. Removing part of the shell may decrease the weld clearance between shell fittings, and the corner weld and modifications may be required to comply with the API 650 spacing requirements. The procedures used to modify shell nozzles are discussed later in this chapter.

When the tank bottom loses integrity or when regulations require leak detection, the most common scenario is to make the existing bottom impermeable, thereby converting the bottom to an RPB. Installing a new bottom by any method will lend itself to

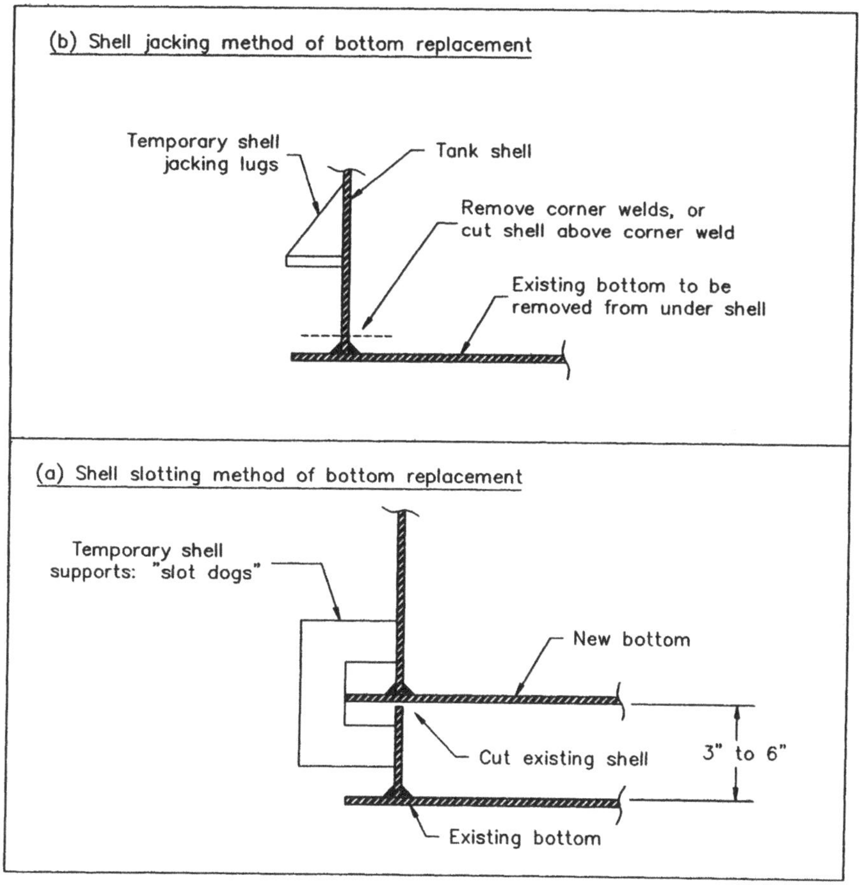

FIGURE 14-1.
Bottom replacement methods.

converting a single bottom tank into a double bottom tank for leak detection, provided that the existing bottom meets the permeability requirements of API 650 Appendix I. This section, as discussed in Chapter 9, requires that the RPB permeability not exceed 1×10^{-7} centimeter per second. Generally, the existing bottom does not meet these requirements because of corrosion or weld defects that may have created holes. If these defects are not detected and the existing bottom is to be used in leak detection, it should be tested to demonstrate permeability. Use of a vacuum box test may be an effective method to check for leaks, not only in the weld seams but also in the plate itself. When the existing bottom does not meet the permeability requirements of API 650 or other government regulations, the most common step is to line the existing bottom with an HDPE liner and install a new primary bottom. This type of double bottom was described in Chapter 13 for new tanks, and the details remain very similar for existing tanks. The liner or RPB must have an acceptable permeability and completely cover the existing bottom, extending up the shell to the new bottom as shown in Figure 9-3. The shell is slotted for the installation of the new bottom, and the space between the bottoms is filled with sand, gravel, or concrete.

Excessive settlement of the bottom, as discussed in API 653 Appendix B, is a common problem that may be revealed during inspection and evaluation of an AST. This appendix outlines acceptable settlement limits and classifies settlement into three different categories: uniform, planar tilt, and out-of-plane or differential settlement. Uniform settlement of a tank occurs homogeneously around the perimeter or across a tank diameter, whereas planar settlement is greater on one side than the other but allows the tank bottom to remain on a given plane. Typically, uniform or planar settlement does not pose any problems for an AST but may for the associated piping, which must be accounted for in the design. The most damaging type of settlement for tanks is differential settlement, defined as a difference in elevation between two points in a given plane, which may be either a horizontal or an inclined plane. Many problems can be associated with differential settlement, such as cracking of the bottom plates, deformation of the floating roof due to excessive settlement of the bottom, and buckling of the tank shell.

Differential settlement of the bottom plates away from the tank shell can be repaired by several methods. First, if the bottom has a number of settlement problems or there exists cracks or severe corrosion, the entire bottom or parts of the bottom may be replaced after the tank is leveled by using sand or gravel. Second, the tank may be converted to a double bottom with the installation of a liner. Again, the bottom first is leveled with sand or gravel, before the liner and primary bottom installation. Finally, if the existing bottom does not require any repairs, pressure grouting could be used in low areas to raise the tank bottom. The appropriate method should be based on the amount and the magnitude of differential settlement or the shape of the existing floor, which may require repair or replacement because of corrosion or cracking.

Differential settlement of the bottom near the shell can result in damage such as shell buckles. Not only does buckling of the tank shell affect appearance, but large buckles may impede floating roof operation or lead to cracking of the plate. The formation of buckles usually is attributed to the differential settlement of the shell or planar settlement of the tank, which may cause the tank to oval. The most common method of removing the shell buckles is to remove the differential settlement by jacking the shell up level and building up the foundation with sand or concrete. If the tank jacking does not remove the buckles, circumferential stiffeners may need to be added to round out the tank shell.

The condition of the shell and the necessary repairs are probably the most critical aspects of AST structural integrity in avoiding a possible spill due to failure. API 653 has dedicated Section 7 to the proper repair or modifications of a tank, and it covers all details from the installation or modification of nozzles to the repair of shell plate. The extent of shell plate replacement is determined from the tank evaluation, whereas new nozzles or other modifications are usually the result of a change of service or operation of the AST.

The amount of each type of shell corrosion present usually dictates the type of modification required for a tank. If the corrosion is uniform across the interior surface of the shell, then the best option may be to decrease the design liquid level. When corrosion occurs in a few random sections of the shell, or the shell exhibits large pitting, the best approach may be to replace the shell plate as required to remove the corroded areas. A common location of corrosion in ASTs storing petroleum is at the water interface. If the water has not been removed consistently over the years of operation, corrosion of this area can be excessive and may require replacement of the first shell plate. It is common for this corrosion area to be within the first three to six inches of the floor. The best approach may be to remove this area of shell plate and jack the tank down so that the shell height will be reduced, or convert a tank that is a single bottom to have an RPB and leak detection.

The most important element in the repair or replacement of shell plate is to ensure that the material grade and thickness match the existing shell plate and conform to the current material requirements of API 650. As discussed in Section 5.3 of API 653, the existing material must be identified by original drawings, API nameplate, or other documentation before any modifications can be conducted. If the material type is unknown, the existing material must be analyzed to confirm the chemical and physical properties, in addition to determining the material toughness by conduction of a Charpy V-notch test. The material thickness should be at least equal to the minimum thickness measurements, but, ideally, should match the as-built thickness or the maximum thickness measured during the inspection.

After the proper material is selected, shell plate installers should avoid current tank shell weld seams as much as possible. This precaution will avoid inducing additional heat in the existing heat-affected zone, which could decrease the material toughness and increase the potential for brittle fracture failure or induce cracking due to stresses from the material cooling. To ensure proper spacing between existing and new weld seams, API 653 (Figure 7-1) specifies the recommended spacing, which is reproduced in Figure 9-5. Generally, for shell plates under one-half inch, new weld seams must be greater than six inches from any existing vertical seams and three inches from horizontal seams. The replacement plate corners should have a six-inch vertical radius and may intersect or remove horizontal seams, provided that the intersection is at approximately 90 degrees and the subsequent new horizontal shell is an adequate distance from the existing seams. Additionally, the standard recommends that the spacing between new shell seams and the edges of shell penetrations must meet the weld spacing requirements of API 650.

API 653 (Section 7) also specifies guidelines for modifying or installing new nozzles, which may be mandated by the AST inspection and completed per general requirements of API 650. New nozzles should be installed according to API 650 or Section 7.7.2. of API 653. Existing nozzle modifications usually are required because of lack of reinforcement or to obtain the minimum weld spacing between the nozzle and the bottom plate when a new bottom is installed. If during the inspection a nozzle is determined to be improperly reinforced or underreinforced, modifications should be completed on the reinforcement according to API 650 and Section 7.6.2 of API 653, which allows the reinforcement to be installed in halves around the shell penetration.

When a new bottom is installed, especially three to six inches above the existing bottom, the weld space between the nozzle weld and the shell-to-bottom corner weld may not meet the requirements of API 650 Section 3.7.3. If this distance is less than the minimums, the shell nozzle reinforcement must be converted to a low-type reinforcing pad, or the nozzle must be raised. The procedure for raising a shell nozzle is described in Section 7.8 of API 653, and consists of cutting the shell around the nozzle with adequate

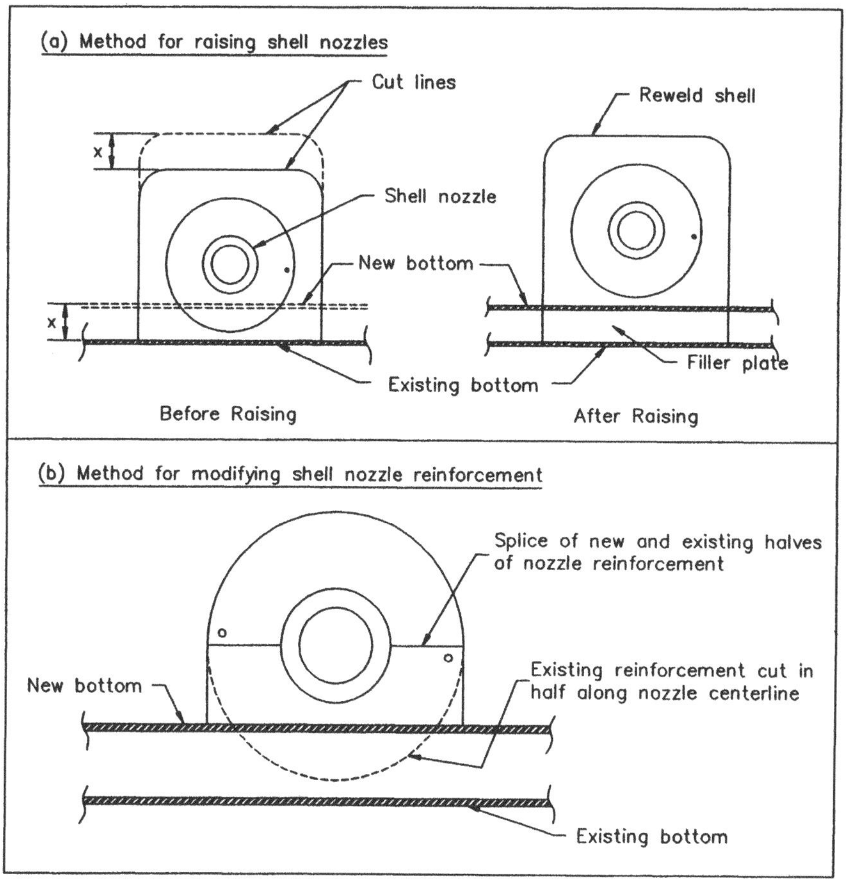

FIGURE 14-2.
Modification of shell nozzles to meet minimum spacing requirements between the corner weld per API 650, Section 3.7.3. (Source: American Petroleum Institute)

clearance, raising the nozzle an amount equal to the space between the two bottoms, and welding the shell plate containing the nozzle. This is pictorially described in API 653 (Figure 7-4) and is reproduced in Figure 14-2(A). Piping considerations could dictate that the raising of existing nozzles is not an option, and would require that the reinforcing pad be converted to a low-type as allowed per API 650. When the reinforcing pad is converted to a low-type, the existing reinforcement is split along the nozzle centerline. The reinforcement added below the nozzle must be checked per API 650 to ensure that the area adequately reinforces the opening and the material used conforms to the standard. This procedure for modifying the reinforcement is given in Figure 14-2(B).

Because of the large liability potential of tank failures, it is important to adequately address all issues uncovered during the inspection and to ensure that subsequent modifications be completed according to the applicable standards, API 650 and API 653. Any modifications also shall be completed by using good engineering judgment and done by a manufacturer competent in the design and the construction of ASTs.

Reconstruction

Often an AST is retired from the service category it was initially constructed for but could be utilized in another part of a facility or at a different terminal, thus requiring the tank to be moved. Reconstructed tanks have been used for a number of years, but it was not until the adoption of API 653 that there existed standards outlining the procedures for reconstructing tanks. API 653 divides the requirements into design considerations, Section 6, and construction procedures, Section 8. The design considerations mainly focus on the adequacy of the tank and the difference in environment the tank will experience at the new location. The tank should undergo a complete evaluation to determine the plate thickness and need for repairs. The most important part of the evaluation is to ensure that the tank shell has adequate material toughness at a new location that may have a lower design metal temperature than the old site. The tank also should be evaluated for a possible change in wind and seismic loadings, which could require stiffening of the tank and anchor bolts if the tank is unstable.

Section 8 of API 653 discusses the procedures for dismantling and reconstructing a tank as well as the dimensional tolerances of the foundation and the tank at the new location. The tank bottom and the roof should be dismantled by deseaming lap weld joints or by cutting two inches away from existing seams, except where cuts cross existing seams. Dismantling of the shell follows nearly the same requirements, with the shell removed by cutting out the existing welds and heat-affected zones or by cutting six inches away from existing seams. According to API 653, the shell-to-bottom corner weld cannot be reused unless the entire bottom is intact. When the corner weld is removed, the shell and the bottom should be cut a minimum of one-half inch from the toe of the corner weld. Reconstructed tanks should be welded in accordance with the applicable standard and by qualified personnel to reduce the potential for a tank failure.

CHAPTER 15

Shop-Built Aboveground Storage Tank Products

This chapter discusses available shop-built aboveground storage tank products. Because of the differing requirements of AST users, manufacturers have developed various designs to capitalize on market trends. Brief discussions of UL vs. API codes and AST venting are followed by sections on the most popular design options for shop-built steel ASTs, divided into four categories: single wall ASTs, double wall ASTs, dike ASTs, and fire-protected ASTs. Finally, a summary table is provided at the end of the chapter to compare and contrast the different products. (Photo 15-1 shows shop-built ASTs at a terminal facility.)

UL vs. API Codes

The most widely recognized standard for shop-built ASTs is UL 142. As previously discussed, UL has set manufacturing guidelines for

PHOTO 15-1.
Shop-built ASTs at a terminal facility. These tanks can be UL 142 or API designed.

shop-built ASTs that are easily identified in the industry (UL 142 and UL 2085). By specifying a UL 142 single wall AST, the tank operator can be assured the AST will be built to fairly exacting requirements regardless of the tank manufacturer selected. The UL brass label is a mark of conformity and security.

Not to be overshadowed, the American Petroleum Institute (API) has two codes that discuss shop-built ASTs: API 650 Appendix J, *Shop-Assembled Storage Tanks,* and API 12F, *Specification for Shop Welded Tanks for Storage of Production Liquids.* These API codes are designed to allow the oil production industry a shop-built AST design from a recognizable authority. In many cases, major oil companies are not familiar with the UL codes, and, more directly, field-erected tank builders do not have licensed UL facilities to manufacture UL ASTs. Therefore, the API codes allow a field-erected tank vendor to supply a smaller shop-built AST manufactured to a recognizable standard.

The main differences between UL and API codes can be summarized in a few words. First, UL has strict manufacturing requirements, whereas the API codes are intended as design guidelines or parameters for an engineer to follow (e.g., "This shop-built AST is built to the UL 142 standard" compared to "This shop-built AST is designed per API 650, Appendix J"). Second, a UL 142 tank can be up to five times less costly than an API shop tank. Many companies quickly switch to UL tanks once they compare the price of equal-sized API and UL ASTs. Also, API does allow the design engineer more freedom in regard to the capacity and the dimensions of the shop-built AST in Appendix J, with a simple requirement that the vertical AST cannot be larger than 20 feet in diameter. Alternatively, API 12F tanks are limited to the following specifications: nominal capacity between 90 and 750 barrels (approx. 3,780 and 31,500 gallons), diameter between 7'11" and 15'6", height between 8 feet and 24 feet. Finally, UL specifies horizontal and vertical configurations, single and double wall designs, dike tanks, and fire-protected tanks. The API codes are for vertical tanks only.

Although API shop tanks serve a function in the shop-built AST market, UL tanks offer a wider range of products than API tanks do.

AST Venting

A major safety concern with all AST products is venting. All ASTs must be properly vented to prevent excessive pressurization or vacuum, which can result in catastrophic failure of the AST system and, consequently, product release. As previously discussed, shop-built ASTs are designed to operate at atmospheric pressure. Any changes in that pressure, such as filling, rapid dispensing, or fire exposure, could cause a dangerous structural problem. Excessive vacuum can cause the AST to implode on itself as the vacuum pulls air out of the AST at a rapid pace, and excessive pressure can cause the AST to explode as air is rapidly expanded in it.

Therefore, normal and emergency vents are required in all AST systems to prevent these occurrences. A normal vent will operate during the normal fill and dispensing cycles to compensate for vapor pressure changes in the atmospheric tank. An emergency vent is intended to relieve buildup of internal pressures that can result from heating of the contents when the tank is exposed to a fire. Usually, the settings for emergency vents are higher than those for normal vents.

Single Wall ASTs

Over the years, single wall ASTs have been recognized as the standard for the storage of all types of liquids. In fact, all tank designs revolve around the single wall primary tank. It would be impossible to travel down any street in America without encounter-

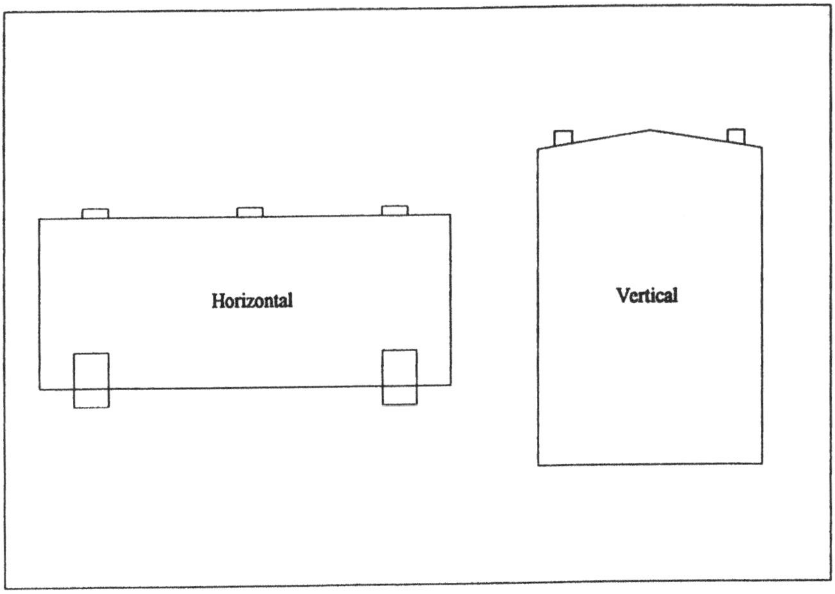

FIGURE 15-1.
Single wall AST design.

ing some form of a single wall storage tank. Traditionally, the single wall tank was the only product demanded and produced by tank manufacturers. Even today, many tank users and operators think that a single wall tank is sufficient for their storage needs. Applicable manufacturing standards are UL 142, API 650 Appendix J, and API 12F.

Single wall ASTs are basic in design and are the most economical type to purchase. Figure 15-1 illustrates two configurations of the single wall design, horizontal and vertical. Horizontal tanks must have steel saddles or anti-roll brackets or must be placed in a concrete cradle to prevent the tank from moving.

Secondary Containment

As all the national regulations and fire codes clearly state, ASTs must have some form of secondary containment to contain accidental spills or leakage. For single wall ASTs, the most popular forms of secondary containment are diking around the tank and placing the tank in a vault.

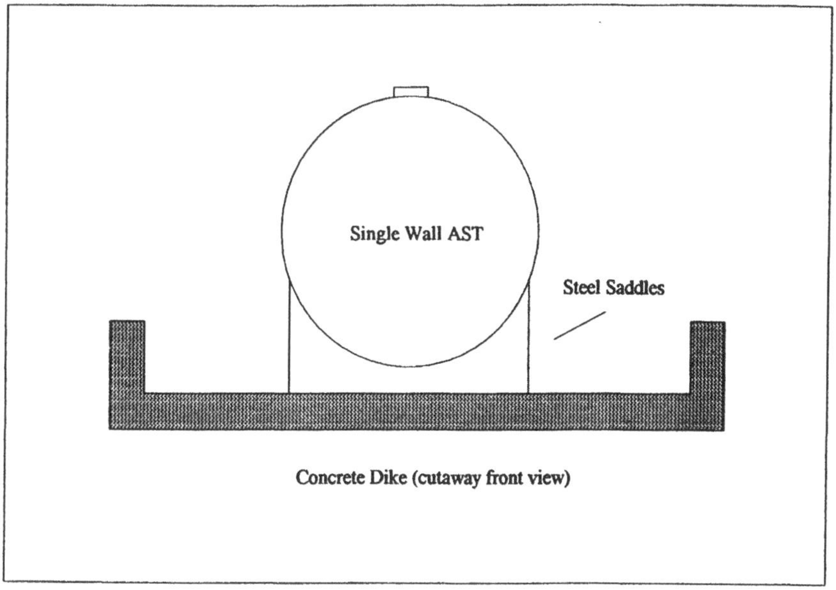

FIGURE 15-2.
Single wall AST in concrete dike.

Diking of a single wall AST can be accomplished in many ways. Some of the materials most frequently used for manufacturing the dike are concrete, asphalt, and earth. Figure 15-2 shows a single wall horizontal AST installed in a concrete dike, and Photo 15-2 shows vertical ASTs installed in a concrete dike. Although concrete and asphalt dikes generally are not internally coated or lined, earthen dikes may be required by the authority having jurisdiction to have an HDPE liner installed. Increasingly, regulators are requiring that dikes meet certain permeability specifications that concrete and/or asphalt may not provide.

Acquisition of a single wall tank would be beneficial in a facility that had several ASTs already installed in a diked area. The facility operator would save a significant amount of money simply by purchasing an additional single wall tank and installing it in the existing diked area. Also, for multiple AST purchases, a facility operator should examine the costs of buying single wall tanks and building a diked area, as opposed to purchasing multiple secondary containment tank technologies.

A fairly recent development (since 1988) in secondary containment is the installation of single wall tanks in vaults. This option was designed as a means of avoiding UST regulations by building a

PHOTO 15-2.
Single wall ASTs installed in a concrete secondary containment dike.
(Source: Brown-Minneapolis Tank)

concrete vault above or below grade and placing an AST inside. Because the tank would not come into contact with the soil, the installation would not fall under UST regulations. A typical concrete vault, with dimensions, is depicted in Figure 15-3. This design would be beneficial for a retail or commercial fueling facility that did not have a large amount of land for multiple ASTs with large capacities, but would like to avoid UST regulations. A facility operator could build below-grade concrete vaults and install single wall ASTs in each vault with larger capacities than might be allowed by fire codes for above-grade fire-protected ASTs.

Only the latest version of NFPA 30A covers the standard design of a vault. Prior to 1994, shop-built ASTs were not allowed at retail

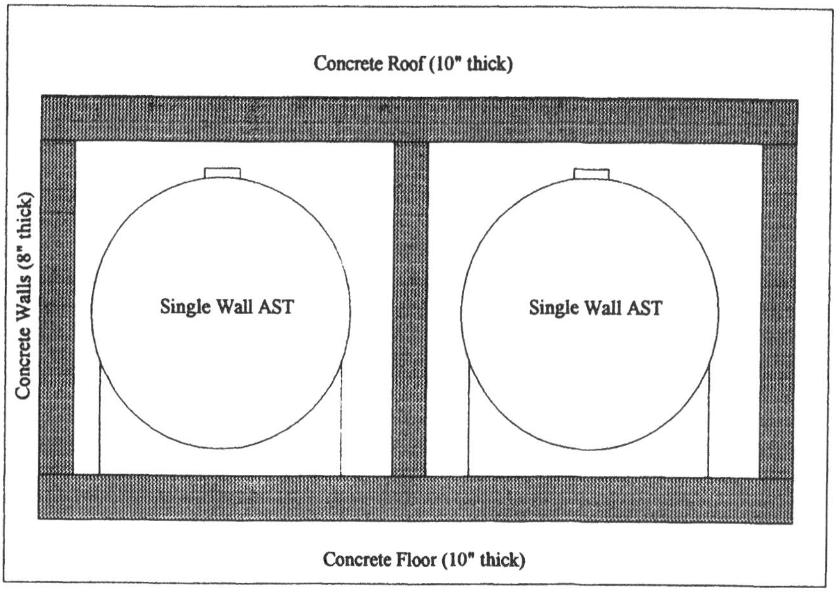

FIGURE 15-3.
Single wall AST in vault.

fueling locations because of potential fire hazards, and fire codes and officials preferred a UST installation. Since that time, UFC and NFPA have recognized acceptable fire-protected tank technologies that make an AST at a retail fueling facility practical; so the enclosed-vault concept does not have as many advantages as it once possessed.

Monitoring the integrity of the bottom of a vertical single wall AST placed directly on the ground can pose potential difficulties. For a vertical single wall AST installed directly on the earth, the steel bottom should be coated with a corrosion-protective coating. In addition, cathodic protection of the tank bottom through the use of sacrificial anodes is a safe and inexpensive means of ensuring bottom integrity. Refer to Chapter 12, STI-R893, *Recommended Practice for Protection of Vertical AST Floors,* to obtain further details on coatings and cathodic protection. Another method of monitoring a vertical-tank bottom is to add a factory-attached secondary bottom. Figure 15-4 illustrates sample options for double bottoms on vertical single wall tanks.

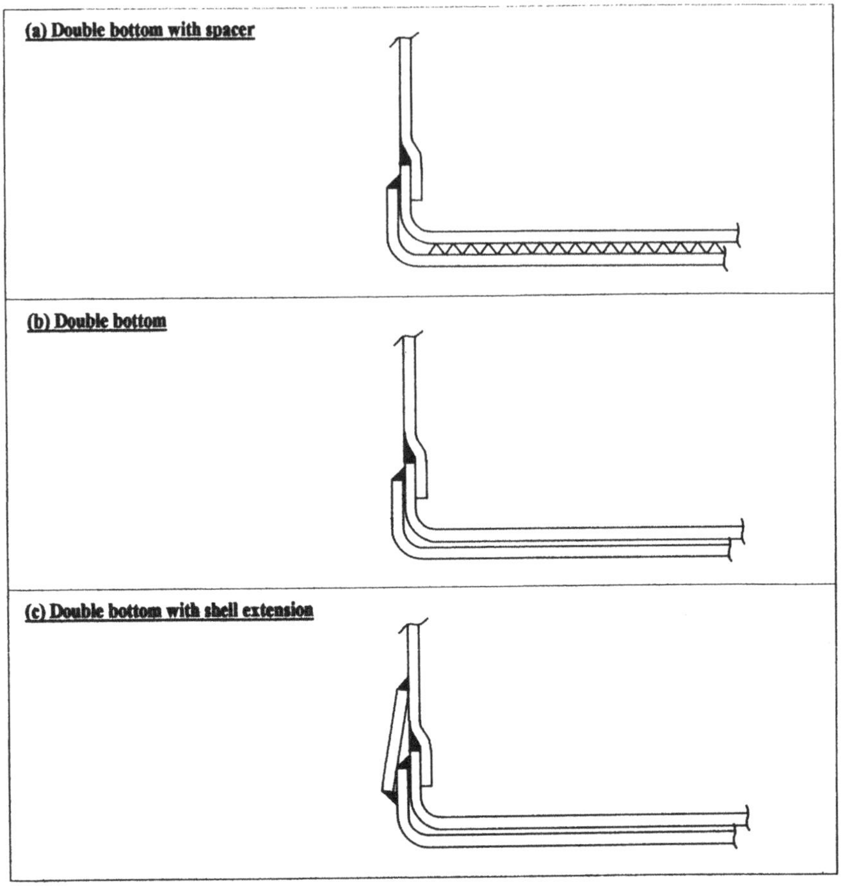

FIGURE 15-4.
Single wall vertical AST with double bottom, three design options.

Double Wall ASTs

The popularity of double wall ASTs has increased greatly since their introduction in 1988. The concept of a double wall tank is not complicated, and, in application, double wall underground tanks have become an industry standard. Taking the double wall UST concept above ground has provided AST operators many advantages. Applicable manufacturing standards are UL 142 and STI F921.

Double wall ASTs are available in horizontal or vertical configurations, as shown in Figure 15-5. The main product meant to be stored in this design is a flammable or combustible liquid. The ad-

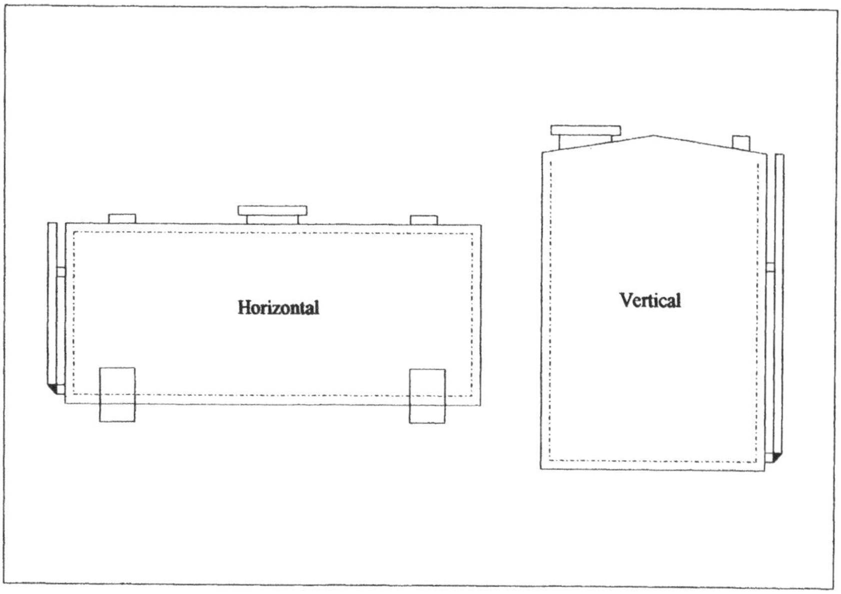

FIGURE 15-5.
Double wall AST (horizontal and vertical).

vantages of the double wall design include: interstitial monitoring, integral secondary containment, limited land use, light weight, and large available capacities.

Interstitial monitoring is accomplished by an external or an internal monitoring pipe or well attached to the interstice between the walls of the double wall tank, as shown in Figure 15-6. Many leak detection devices, from simple mechanical float devices to sophisticated electronic instruments, are available that will fit into the monitoring well.

Integral secondary containment forming an interstitial space between the walls of the double wall AST is a major advantage of this design. In the event of a leak in the primary tank, the 360-degree secondary containment will contain the leak without any liquid leaving the system, as visualized in Figure 15-7. The standard double wall design will offer 100% containment of the primary tank (Type I design). If desired, the secondary head of the tank can be extended to offer 110% containment or more (Type II design). Overfill protection equipment should be installed in this design, as it is intended for use as an integral secondary containment system.

Owing to the integral design of the double wall AST, this system will take up about the same amount of land as a single wall AST.

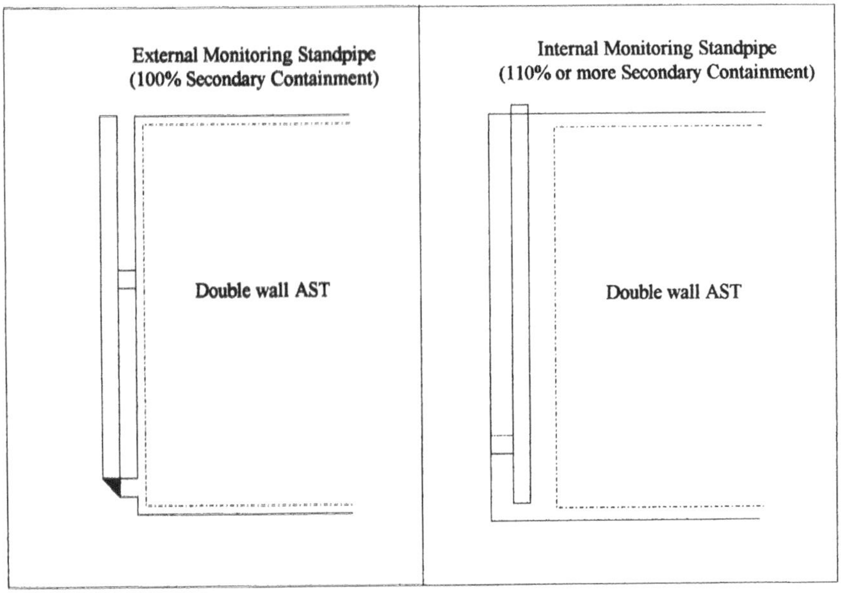

FIGURE 15-6.
Monitoring—external and internal.

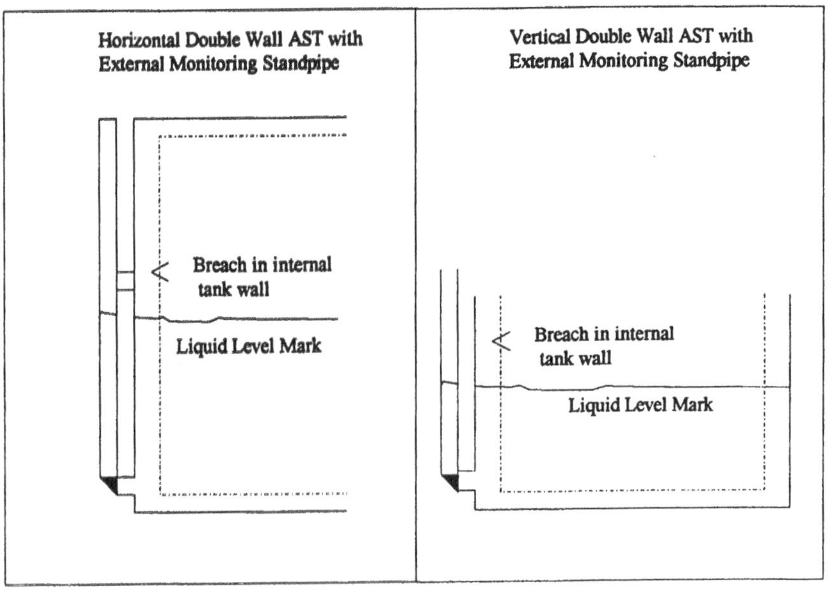

FIGURE 15-7.
Horizontal and vertical double wall ASTs with leak detection.

When one is deciding on the appropriate AST product for a facility, inevitably the issue of available space must be resolved. With a double wall AST, no additional diking would be required; so this system would leave the smallest footprint of all the ASTs.

Because the double wall design is, in essence, a tank within a tank, it is lightweight compared to a dike tank or a fire-protected tank. This lightness leads to easier installation and reduced shipping costs for the entire system.

One major benefit of the double wall design is the large capacities available for various storage needs. A horizontal double wall design can be built, by certain steel tank manufacturers, in capacities of up to 50,000 gallons. As previously stated, the fire codes do not allow tank capacities in excess of 12,000 gallons for fuel dispensing into motor vehicles. For an industrial company that requires the storage of liquid products (jet fuel, chemicals, fuel oil, etc.) that will not be dispensed into a motor vehicle, a double wall AST can be the solution. Moreover, a double wall tank can be compartmentalized to offer multi-product storage in a single unit.

Photo 15-3 shows a double wall AST installed at a fleet fueling facility.

Diked ASTs

The concept of a single wall primary tank placed in a secondary containment dike is fairly simple. However, if that secondary containment dike is shop-assembled to the single wall primary tank and shipped as a complete system, then it is a viable AST product. Today, the unitized dike tank concept is very popular around the United States, and various unique designs have been developed. This section highlights the two options for diked ASTs: the open-top diked AST and the closed-top diked AST.

Open-Top Diked AST

The open-top diked AST is the least expensive and most uncomplicated of the diked AST line. This design is adequate for many tank operators and allows secondary containment, visual leak inspection, overfill or spill containment, and ease of liquid recovery in the diked area. Capacities for this design are limited to about 20,000 gallons because of shipping constraints. Figure 15-8 shows an open-top

PHOTO 15-3.
Double wall AST installed at a fleet fueling facility with attached dispenser. (Source: Brown-Minneapolis Tank)

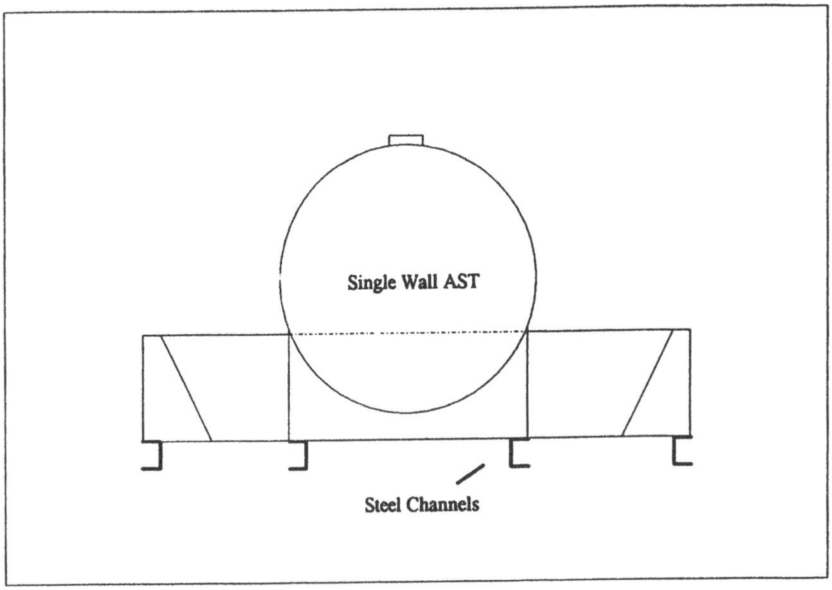

FIGURE 15-8.
Open-top diked AST.

PHOTO 15-4.
Open-top diked AST. (Source: Steel Tank Association)

diked AST allowing for 110% containment of the primary tank. (See also Photo 15-4.) The two national standards for open-top diked ASTs are UL 142 and STI F911.

The main issue with this design is the concept of the open-top dike. The benefits are these: visual inspection of the primary tank is easy, any overflow from valves or fittings will fall directly into the diked area, the dike will protect the tank from vehicles or machinery, and protection of the primary tank from a nearby fire can be accomplished by filling the dike with foam. There also are disadvantages of this design: during inclement weather, rain or snow will fill up the diked area, thus reducing the containment potential, and the water will be considered contaminated; and debris can be discarded into the diked area. Therefore, personnel will have to clean the area on a regular basis, thus increasing maintenance costs.

Closed-Top Diked AST

The closed-top diked AST was developed because of problems with debris entering the diked area. The addition of a so-called rain shield to the design would appear to solve any problem with the diked AST

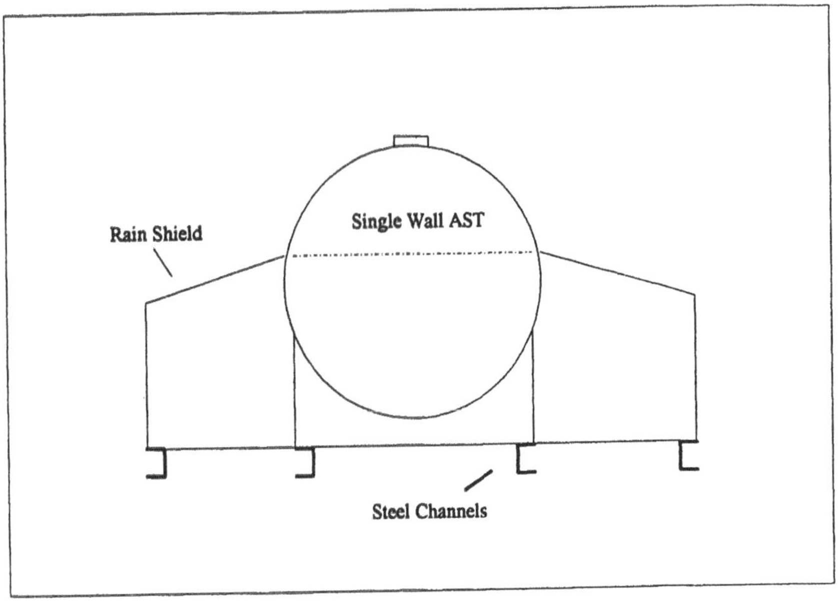

FIGURE 15-9.
Closed-top diked AST.

system. Capacities for this design are limited to about 20,000 gallons owing to shipping constraints. Figure 15-9 portrays a closed-top diked AST allowing for 110% containment of the primary tank. The applicable manufacturing standard is UL 142.

Benefits of the closed-top system include: secondary containment, a debris-free containment area, vandalism protection, and vehicular/machinery protection of the primary tank. There are several disadvantages of the closed-top design: the diked area could have excess vapor buildup that would require an emergency vent; it is difficult to visually inspect the diked area; access to the diked area is limited by the closed top; overfill or spill devices are required, as the rain shield will prevent liquid from entering the dike; and firefighting is inhibited by the large closed-top dike, which prevents personnel from reaching the tank from a safe distance.

Because the closed-top diked AST requires additional equipment for proper operation, it is more costly than the open-top design. Also, the only national standard for closed-top diked ASTs is UL 142, which does not give many specific details of this design; so, in most cases, the AST operator must rely on the individual manufacturer's

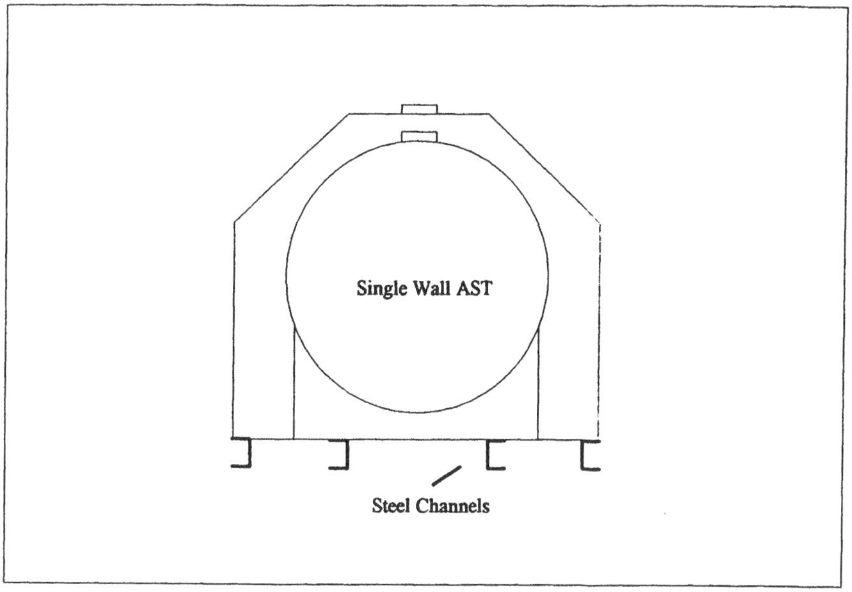

FIGURE 15-10.
Fully enclosed diked AST.

design interpretation of the UL 142 code, which will vary by manufacturer.

Another option with a closed-top diked AST is to fully enclose the dike area, a relatively new method. The design completely encompasses the primary tank in a truly closed-top system. Capacities for this design are limited to about 20,000 gallons owing to shipping constraints. Figure 15-10 portrays a fully enclosed diked AST allowing for 110% containment of the primary tank. The disadvantages associated with the closed-top design of Figure 15-9 are enhanced by this design. Certain manufacturers marketing this design will sell it as a complete unit with pumps, fittings, and accessories. This would be the most expensive of the diked AST systems, but it would include all the necessary components.

Fire-Protected ASTs

Based on their acceptance by various fire authorities, fire-protected ASTs have become the preferred AST product for numerous motor vehicle fueling installations. This design has allowed aboveground storage tanks to be installed at motor vehicle fueling locations. Although fire authorities would prefer a UST installation, they consider the new generation of fire-protected ASTs to offer a comparable degree of safety. Applicable manufacturing standards are UL 2085 and STI F941.

Since the inception of fire-protected ASTs, various designers have tried to improve on the concept and eliminate its disadvantages. Unfortunately, exorbitant system costs and weight constraints cannot be totally eliminated. The basic premise of a fire-protected AST is that a steel primary tank is surrounded by six inches of concrete or insulation, which provides an adequate level of fire resistance. The various designs for this system can be divided into two categories: single wall with fire protection and double wall with fire protection.

The original fire-protected AST, which consisted of a single wall rectangular primary tank surrounded by six inches of concrete, usually was limited to about a 2,000-gallon capacity. (*Note:* Today the maximum capacity is 10,000 or 12,000 gallons.) The rectangular design allowed a dispenser to be installed on top of the unit but still to be in reach of personnel to fuel motor vehicles. The issue of secondary containment for single wall fire-protected ASTs is of particular interest. Although construction methods vary by manufacturer, one way of containing the primary tank is to enclose it in a membrane and then have it encased in six inches of concrete. Single wall fire-protected ASTs can be shipped from the factory as a complete vessel or can be shipped separately and assembled on-site, either by pouring the concrete on-site or by connecting two halves of the concrete vault assembly. Figure 15-11 shows a rectangular single wall fire-protected AST. These single wall designs can still meet UL 2085 for an "insulated aboveground tank," but they cannot bear the UL 2085 label for an "insulated secondary containment aboveground tank." Also, UL 2085 specifically states that in order to be a UL 2085 *listed* system, the entire assembly must be shipped from the factory as a single unit.

Today, the preferred method of many AST operators is to install double wall fire-protected ASTs. The double wall design is available in rectangular or cylindrical configurations. The rectangular design usually is available in sizes up to 12,000 gallons, with the cylindrical design available in sizes up to 30,000 gallons. The basic design

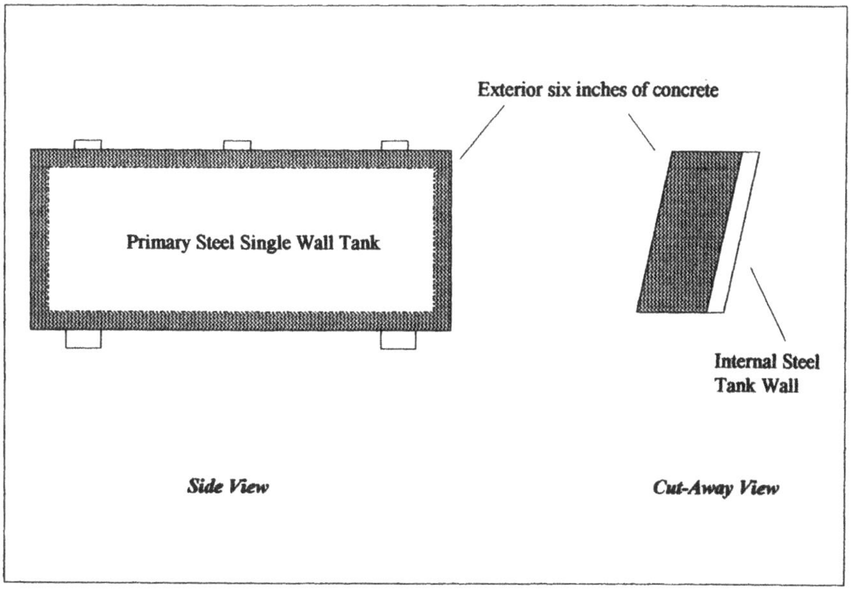

FIGURE 15-11.
Single wall fire-protected AST (rectangular).

consists of a primary tank and a secondary tank separated by an interstitial space of six inches that is filled with a fire-resistant insulating material. Leak detection is accomplished by a monitoring pipe installed into the interstitial space. Another option would be a UL 142 double wall tank surrounded by six inches of concrete. (*Note:* This option may have an outer steel wall protecting the concrete, which would make this a triple containment system.) Figure 15-12 illustrates a rectangular double wall fire-protected AST, and Figure 15-13 shows a cylindrical double wall fire-protected AST.

With all the options available in fire-protected tanks, here are a few questions to consider in making a purchasing decision:

- Does the insulating material consist of heavy concrete or light insulation? Light concrete without gravel can weigh up to 75% less than heavier concrete, with an equivalent degree of fire protection. This can lead to savings in shipping and installation costs.
- What is the warranty? Many of the fire-protected AST designs come with extended warranties. Some manufacturers offer a 20-year standard warranty with an optional 30-year warranty at additional

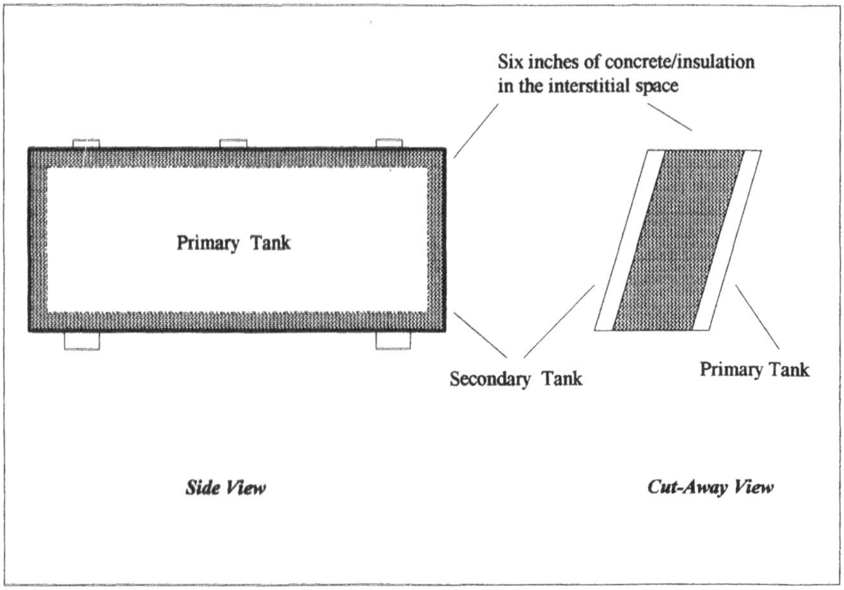

FIGURE 15-12.
Double wall fire-protected AST (rectangular).

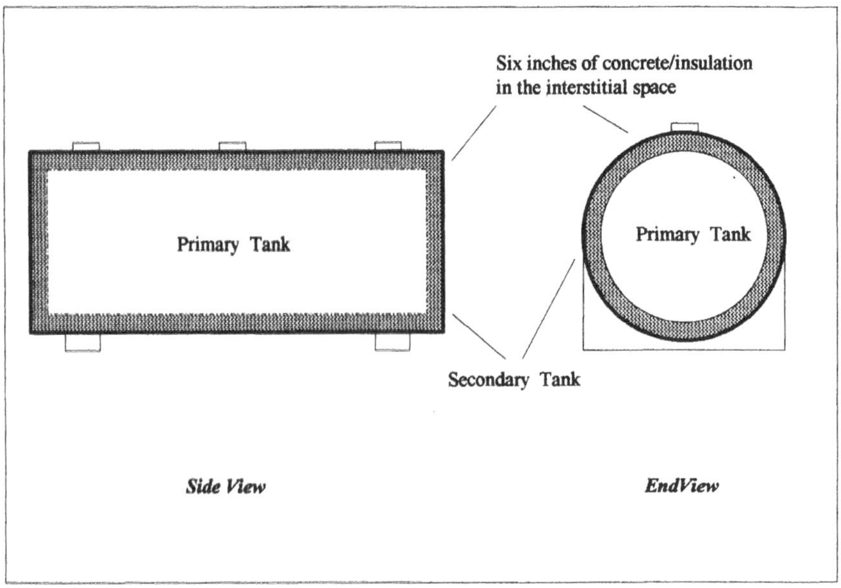

FIGURE 15-13.
Double wall fire-protected AST (cylindrical).

expense. It is important to determine if the warranty is backed up by an independent insurance company or the manufacturer. A warranty from the manufacturer is only as good as that manufacturer's financial stability.

- Does the tank system have a UL 2085 listing for an "insulated tank" or "an insulated secondary containment tank"? Will the "insulated tank" require a secondary containment dike?
- Does the tank system need to be installed on a concrete pad? (Of what size and thickness?) Will an existing concrete pad at the facility in which the tank is to be installed be sufficient?
- Does the tank system include spill prevention, overfill protection, leak detection, and a manway for future access? Also, is emergency and normal venting supplied with the tank system? Does the system include a dispenser? When comparing competing systems, one must consider all the auxiliary equipment that is included.
- Does the tank system ship with a finished exterior coat of paint or another protective coating? Some manufacturers include only an exterior coat of shop primer in the standard price.
- What fire codes does your local county/city recognize, NFPA or UFC? Will the tank system meet the applicable fire code?
- If your local county/city fire code is UFC, will the fire authority require the AST system to have passed the optional UFC Appendix Standard A-II-F-1 projectile protection test? This test requires that without penetration of the primary tank, the protected tank assembly must resist five rounds of 150 grain M-2 ball ammunition, having a nominal muzzle velocity of 2700 feet per second fired from a .30 caliber rifle at a distance of 100 feet.
- For nonretail motor fuel dispensing (i.e., industrial facilities, commercial facilities, government, etc.), what will be the maximum allowable capacity per tank? In some cases, the local fire authority will have to grant a variance to exceed the NFPA or UFC capacity requirements. From an economic standpoint, it will be less costly to buy a few larger-capacity tanks than numerous small-capacity tanks. (*Example:* An industrial facility requires 40,000 gallons' worth of storage capacity. One option is to have four 10,000-gallon ASTs installed. Another option is to install two 20,000-gallon ASTs. Installing two 20,000 -gallon ASTs will be substantially cheaper in tank cost and installation cost, and two tanks will take up less real estate than four tanks.)

Summary

To conclude this discussion of AST product choices, it should be noted that the AST purchaser must weigh many factors. To make the proper decision, one must consider all available options. Table 15-1 highlights the major aspects of each tank technology for such comparison.

Table 15-1. AST product choices for fuel dispensing

Tank design	Liquids stored	Mfgr. codes	Advantages	Disadvantages
Single wall AST	All	UL 142, API App. J, API 12F	Is least expensive, is easily relocated, meets NFPA with additional spacing, is lightweight, is vertical or horizontal, has large capacities to 50,000 gallons, is practical and functional, is easy to install	Does not meet UFC/BOCA/SBCCI fire codes, must provide secondary containment, has no monitoring capability, requires saddles or concrete cradle, usually has one-year warranty
Double wall AST	Flammable and combustible	UL 142, STI F921	Meets NFPA with additional spacing, is easily relocated, has built-in monitoring, has integral secondary containment, is lightweight, is vertical or horizontal, has large capacities to 50,000 gallons, has industrial applications, takes same space as single wall for installation	Does not meet UFC/BOCA/SBCCI fire codes, requires saddles or concrete cradle, usually has one-year warranty
Diked AST	Flammable and combustible	UL 142, STI F911	Meets NFPA with additional spacing, is easily relocated, has built-in secondary containment, has rectangular design, has capacities to 30,000 gallons, has industrial applications	Does not meet UFC/BOCA/SBCCI fire codes, has heavy and cumbersome rectangular steel dike, requires large footprint of land for installation, has high shipping costs, with open-top diked area is easily contaminated by weather or debris, usually has one-year warranty
Fire-protected AST	Flammable and combustible	UL 2085, STI F941	Meets NFPA/UFC/BOCA/SBCCI, is horizontal or rectangular, has built-in monitoring, has built-in secondary containment, usually has 20–30-year warranty	Most expensive, has excessive weight, requires concrete pad under system, has limited capacities, is mainly for vehicle fueling, has high shipping and installation costs

CHAPTER 16

Shop-Built Aboveground Storage Tank Installation Instructions

For safe and successful AST installation, it is important to learn the proper testing and installation techniques available in the industry. Although all ASTs should come with installation and testing instructions from the manufacturer, this chapter highlights the main issues for one to consider when performing an AST installation.

There are very accurate and reputable national sources available for shop-built AST installation instructions. First, the Steel Tank Institute (STI) Standard R912-91, *Installation Instructions for Factory Fabricated Aboveground Tanks,* was last revised in April 1991 and focuses on single wall ASTs. Also, the Steel Tank Institute Standard R931-93, *Double Wall Aboveground Steel Storage Tanks Installation and Testing Instructions,* was published in June 1993 and covers double wall ASTs. The main focus of the R931 publication is the testing of horizontal and vertical double wall tanks. For fire-resistant ASTs, one should refer to the Steel Tank Institute

Standard R942-94, *Fireguard™ Installation and Testing Instructions,* published in February 1994. In addition, the Petroleum Equipment Institute (PEI) Standard RP200-92, *Recommended Practices for Installation of Aboveground Storage Systems for Motor Vehicle Fueling,* was published in 1992 and covers both single and double wall ASTs in a single publication.

This chapter combines the major elements of these individual shop-built AST installation and testing instructions in a concise format.

Foundation and Site Preparation

These instructions are intended for stationary, shop-built ASTs that will be storing petroleum products. The foundation must be designed to support the tank and 100% of the weight of the maximum amount of product the tank will be storing. The foundation can consist of concrete, asphalt, gravel, or other stable bedding material that prevents movement, rolling, or uneven settling of the tank, and is designed to minimize corrosion of the tank bottom. A cylindrical, horizontal double wall AST on saddles may be placed on concrete footings that will support the weight of the system.

One should refer to applicable National Association of Corrosion Engineers (NACE) standards and the Steel Tank Institute R893-89, *Recommended Practice for External Corrosion Protection of Shop Fabricated Aboveground Tank Floors* (Chapter 12), about corrosion protection concerns if the tank will come in contact with the ground. As discussed in previous chapters, corrosion protection of the AST is accomplished through the application of dielectric coatings, electrical isolation, and supplemental cathodic protection systems (magnesium anodes). Also, one should consult NFPA 30/30A, UFC, and the local fire marshal for restrictions such as spacing from buildings, property lines, and so on.

The location of ASTs should allow for traffic flow patterns, product deliveries, dispensing islands, and miscellaneous equipment such as lighting or monitoring systems. Most ASTs should be protected by fencing or vehicular barriers—local fire or building codes might specify standard requirements. Certain installations, those close to navigable waterways, will be required to prepare a federal EPA SPCC Plan (refer to Chapter 3 for SPCC details).

Tank Handling

It is important for one not to handle or install a tank without having knowledge of and experience in the procedures involved with proper and safe installation of an aboveground tank for storage of flammable and combustible liquids. Reliance on skilled, licensed professional installers is of major importance, to avoid tank damage and system failures.

The tank should not be dropped or dragged, but should be carefully handled by using cables or chains of adequate length and size attached to the tank lifting lugs. The use of nylon straps and padding can prevent damage to the tank coating. As a point of safety, one must not handle or move the tank unless it is empty of any substance. These ASTs are intended to be used as stationary tanks and should not be used for the transportation of any product.

Finally, the AST should be checked for damage to the surface coating. If any bare steel is exposed, one should apply additional coating to such areas, which should be supplied by the manufacturer in a touch-up kit.

Photo 16-1 shows double wall ASTs being installed.

PHOTO 16-1.
Double wall ASTs during installation. (Source: Brown-Minneapolis Tank).

AST Testing

Testing the integrity of a single wall or double wall AST at the installation site is necessary to ensure that all weld seams remained intact during transportation. Air pressure testing is the most popular method although vacuum testing of the interstitial space of a double wall AST is an acceptable alternative. This section describes the testing procedures.

Testing Preparation, Single and Double Wall Tanks

To prepare for the air test, remove all thread protectors and install metal pipe plugs. Remove emergency vents and install metal plugs in the tank vent threaded opening. If a tank is equipped with a long-bolt manway for emergency venting, replace the manway long bolts with short bolts and tighten them securely. Use only calibrated diaphragm-type air pressure gauges with a zero to ten (0.0–10.0) psig dial span. Aboveground vertical tanks may have a "weak shell-to-roof" seam; so stay clear of the tank top and fittings when pressurizing the tank. Also, do not leave a pressurized tank unattended, and do not stand in front of tank heads or fittings when pressurizing the tank. Finally, do not apply a vacuum to the primary tank of a double wall AST or to a single wall tank.

Testing of Single Wall Tanks, Horizontal and Vertical

For single wall tanks, the internal tank air test pressure should not exceed 1.5 psig for vertical tanks or 5.0 psig for horizontal tanks. Apply a soap solution to all exterior tank surfaces, including all weld seams and fittings, while the tank is pressurized to test for leaks. The presence of bubbles indicates a potential pinhole leak in the weld seam or fitting that should be repaired and retested.

Testing of Double Wall Tanks, Horizontal and Vertical

For double wall tanks, air-pressure-test the primary tank only, or vacuum-test the interstitial space as an alternative. Vacuum testing of the interstitial space of a dual-wall tank can be used instead of air pressure testing of the primary tank. Some manufacturers may ship the double wall AST with a vacuum maintained in the interstitial space. If so, on-site air testing would be unnecessary if the vacuum were maintained within the manufacturer's guidelines. (*Note:* UL 142 specifies applying a vacuum to the interstitial space of at least 13 inches of mercury for 12 hours. If the tank can maintain the mercury level (±2 inches of mercury) for the required time, the test is successful. Also, air pressure testing does not apply to double wall tanks equipped with interstitial vacuum monitoring systems intended to be used for leak detection.)

Install test piping as shown in Figure 16-1 for horizontal tanks and Figure 16-2 for vertical tanks. (*Caution:* For horizontal ASTs, the regulated air supply for this test is not to be less than 3.0 psig or more than 5.0 psig. For vertical ASTs, the regulated air supply for this test is not to be less than 1.5 psig or more than 2.5 psig.) For horizontal ASTs, set the pressure relief valve in the test air supply line at 4.5 psi. For vertical ASTs, set the pressure relief valve in the test air supply line at 2.0 psi. Refer to Table 16-1 for a comparison of horizontal and vertical air requirements.

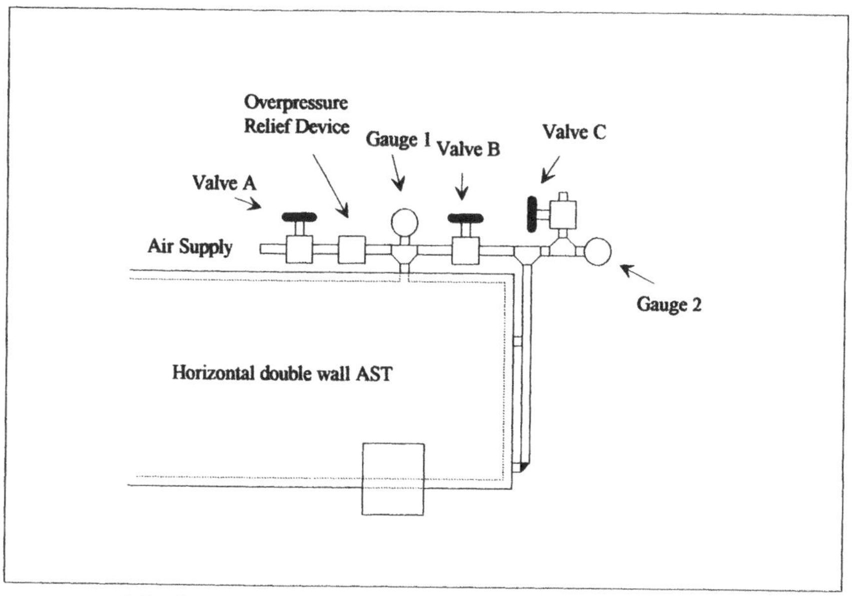

FIGURE 16-1.
Air pressure test of double wall horizontal AST.

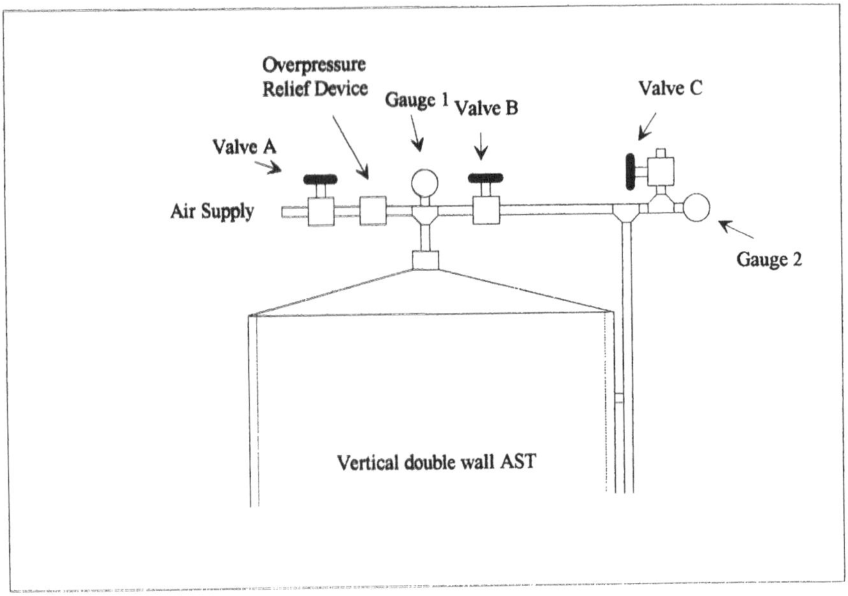

FIGURE 16-2.
Air pressure test of double wall vertical AST.

Table 16-1. Air test pressure requirements

	Horizontal double wall ASTs	Vertical double wall ASTs
Regulated air supply	3.0–5.0 psig	1.5–2.5 psig
Pressure relief valve setting	4.5 psi	2.0 psi

To conduct the air test, refer to Figures 16-1 and 16-2. First, close valves A and B. Then, open valve C. Connect the regulated air supply line to test piping as shown in Figures 16-1 and 16-2. (*Caution:* Determine that the regulated test air supply line pressure is properly set before proceeding.) Next, slowly open valve A to pressurize the primary tank. For horizontal ASTs, pressure gauge 1 should indicate a test air pressure of 3.0 psig to 5.0 psig. For vertical ASTs, pressure gauge 1 should indicate a test air pressure of 1.5 psig to 2.5 psig. Next close valve A, and then disconnect the regulated air supply line from the test piping. Hold the test pressure in the primary tank for one hour minimum. A steady drop in the pressure reading for gauge 1 indicates there may be a leak in the primary tank.

If no leaks are found, then close valve C and slowly open valve B to pressurize the interstitial space between the double walls of the tank. (*Warning:* Do not apply air pressure to the interstitial space between the walls of a double wall tank without air pressure in the primary tank, and do not apply air pressure in the interstitial space that is higher than the air pressure in the primary tank. Damage to the tank may result. (*Note:* Having a greater pressure in the interstitial space than inside the primary tank will result in an external pressure on the primary tank. In general, ASTs cannot withstand a sizable external pressure greater than 1.0 oz/IN^2, which could result in buckling failure of the shell.) For vertical ASTs, pressurization of large-diameter vertical tanks may result in bulging of the bottom of the tank. Should the bottom of the tank move, immediately relieve the pressure. In large-diameter vertical tanks it is possible to lift the tank by pressurizing the inner tank and the space between the bottoms. Therefore, secure the tank accordingly before testing.

As the interstitial space fills with air, gauge 1 will indicate a slight drop in test pressure when valve B is opened, but should hold steady at the lower pressure. If the test pressure drops below the minimum requirements of 3.0 psig (horizontal) or 1.5 psig (vertical), close valve B, reconnect the air supply line, and slowly open valve A to increase the pressure in the primary tank to between 3.0 and 5.0 psig (horizontal) or 1.5 and 2.5 psig (vertical). When the required pressure is indicated on gauge 1, close valve A and disconnect the test supply line. Open valve B to equalize the pressure in the primary tank and in the interstitial space (i.e., gauge 1 and gauge 2 should have the same pressure reading). Close valve B, and then hold the test pressure for one hour minimum in the interstitial space. A steady drop in pressure for gauge 2 indicates there may be a leak in the interstitial space. Immediately apply an approved leak test solution (soap) to tank exterior surfaces, welds, fittings, and so forth. Check for leaks (i.e., soap bubbles around fittings). If leaks are found, notify the tank manufacturer. If no leaks are found, the testing of the tank is complete. To finish the test, open valve B and then slowly open valve A to release the test air pressure from the primary tank and the interstitial space at the same time.

With the tank depressurized, remove all test piping, temporary plugs, caps, and seals. Then, reinstall emergency relief vents, and so forth, or, if the tank is equipped with an emergency vent long-bolt manway, reassemble the manway cover with bolts supplied by the tank manufacturer. (*Warning:* Remove the temporary manway cover short bolts, and reassemble the manway cover with the proper size and type of long bolt, with a minimum unthreaded length of 2.0 inches, so that when the cover is fully raised, there is a 1.5-inch minimum space between the manway cover and the frame. Failure to properly assemble the cover of a long-bolt manway used for emergency venting may make the vent inoperable, causing tank failure by overpressurization.)

Installed Tanks—General Considerations

All UL 142 listed ASTs (both the primary tank and the secondary containment) must have emergency and normal atmospheric vents. Tanks designed for underground use (i.e., UL 58 or UL 1746) may not be used aboveground. Also, all permanent piping and fittings should be installed with compatible, nonhardening thread sealant material. All unused tank openings must be properly sealed by using metal threaded pipe plugs, flanges, or caps with compatible nonhardening thread sealant material. Also, one should not modify the tank structure or weld on the tank. Installed tanks should be properly marked and labeled. Note that different tanks designs often require different labels and markings.

Secondary Containment

Secondary containment around aboveground tanks and the associated piping is designed to contain product spills and releases from tanks, piping, and associated equipment. Secondary containment of aboveground tanks is a federal requirement. The forms of secondary containment are numerous and can be an integral part of the tank, such as double wall tanks or factory-attached steel dikes. When steel secondary containment is employed (i.e., a double wall tank), it is recommended that a suitable vehicle barrier be provided. One method would be to use four-inch metal pipes filled with concrete installed at least two feet away from the tank. The pipe should be a minimum of 42 inches below ground level and 36 inches above ground level.

Traditional dike walls and bottoms for single wall tanks can be constructed of earth, steel, concrete, or solid masonry. The dike walls and bottom must form an impervious barrier to liquid; so earthen dikes should be topped off by a membrane liner, and concrete dikes should be coated. The dike capacity must be at least equivalent to the greatest amount of liquid that can be released from the largest full tank within the diked area, but local jurisdictions may require additional capacity. The Petroleum Equipment Institute dike sizing calculations are presented in Appendix F of this book.

Another method of secondary containment is to install the AST in a vault, which is an essentially liquid-tight structure that completely surrounds the storage tank. By definition, special enclosures are liquid-tight, not back-filled, and constructed of at least six inches of reinforced concrete. The use of special enclosures usually is limited to situations where buildings or property limitations make installations of diked ASTs impractical. The permission of local fire authorities generally is required before special enclosures may be installed. Vaults should completely enclose each tank, and each tank should have a separate vault although adjacent vaults may share a common wall. Tanks designed for underground use may not be used in vaults; only listed ASTs can be used in a vault. Vaults may be above or below ground and should be constructed with a minimum of six inches of reinforced concrete. Vault roofs for an aboveground vault should be constructed of a weaker material than the walls so that in the event of an explosion the thrust of the explosion will be directed upward. For a vault located underground, the roof may be constructed with sufficient strength to serve as a component of a driveway. Vault openings designed for personal entry should be liquid-tight and equipped with a means of locking. Additionally, openings may be used for filling and dispensing operations. Normal tank vents should extend outside the vault, terminate not less than twelve feet above the general grade level, and be located at least five feet from building openings. An alarm leak detection system is recommended for a vault. For an above-grade vault, collision protection should be considered.

Pumps and Valves

Pumps and valves should be designed and fabricated from suitable materials, with adequate strength and durability to withstand the operating pressures, structural stress, and exposures to which they may be subjected. Dispensing from ASTs should not be accomplished by pressurizing the tank or by gravity flow. Two suitable types of pumping systems are suction pumping systems and remote pumping systems. Suction pumping systems typically employ a suction pump mounted on a pump island located away from the AST. In remote pumping systems, the supply piping, between dispenser islands and remote pumps in or near aboveground tanks, operates under pressure. A fire valve should be installed at any tank opening through which stored product could flow by gravity. A piping system should contain a sufficient number of lockable block valves to allow the tank and the piping to be isolated and secured. Where a tank is at an elevation that produces a gravity head on the dispensing unit, release of the tank contents is possible through a leak in the piping or the equipment. An anti-siphon valve should be installed to prevent gravity/discharge flow into the discharge piping when the dispenser is not in use. Check valves prevent the backflow of liquid and should be installed on tank fill connections below the liquid level. Any segment of piping that could be subjected to excessive pressure should be provided with pressure relief valves.

Tank Attachments

Tank attachments should be fabricated from materials with adequate strength and durability to withstand the operating pressures, structural stress, and exposures to which they may be subjected. Tanks must be protected from unacceptable pressure or vacuum during filling or emptying operations, and when subjected to outdoor ambient conditions. Thermal expansion and contraction factors are significantly greater in ASTs than in USTs. During tank fabrication, a connection is provided for each tank opening by welding a steel pipe coupling or threaded flange to the tank to allow the installation of ancillary equipment (normal vent, emergency vent, manway, water draw-off, etc.). The filling of ASTs from transport vehicle tanks normally is accomplished with a pump, either mounted on the vehicle or permanently fixed in the tank-filling area. Provisions should be made for accidental spill prevention and overfill protection by means of an overfill warning or shutdown device. Normal operating pressures for ASTs are limited to service at 1.0 psig maximum internal operating pressure and to 2.5 psig under emergency venting conditions. Allowable vacuum should not exceed 1.0 ounce per square inch. (*Warning:* For tanks tested at 5.0 psig pressure, normal vents should be set to open at less than 1.0 psig and not more than 1.0 ounce of vacuum.) Emergency venting is provided for vapor release in the vent of excessive pressure buildup resulting from fire exposure of the tank.

Pipe and Fittings

An efficient piping design minimizes distances, enhances operations, and facilitates testing and maintenance. Piping must have adequate capacity, strength, and durability to withstand the operating pressures, structural stress, and exposure to which it may be subjected during construction and subsequent operations. Acceptable material selection and installation practices are prescribed in the American National Standards Institute code for pressure piping, ANSI B31.

Appendixes

APPENDIX **A**

Environmental Regulatory Offices

This appendix is divided into two sections. The first section lists contact information for environmental regulatory offices in the United States, including the federal EPA by region, other federal agencies, federal environmental hotlines, and state and territorial regulatory authorities. The second section provides contact information for environmental regulatory offices in Canada, for Environment Canada and for the provincial regulatory authorities.

Environmental Regulatory Offices—United States

Federal

Environmental Protection Agency Headquarters
401 M St. SW
Washington, DC 20460
(202) 382-7963

Table A-1. United States EPA regional offices

Region	Address	Mailcode	City/State	Telephone
U.S. EPA, Region 1	JFK Federal Bldg.	HPU-1	Boston, MA 02203	(617) 573-9604
U.S. EPA, Region 2	26 Federal Plaza	2AWM-HWPB	New York, NY 10278	(212) 264-3384
U.S. EPA, Region 3	841 Chestnut St. Bld.	3HW63	Philadelphia, PA 19107	(215) 597-7354
U.S. EPA, Region 4	345 Courtland St. NE	4WM-GP	Atlanta, GA 30365	(404) 347-3866
U.S. EPA, Region 5	230 S. Dearborn St.	5HR-JCK-13	Chicago, IL 60604	(312) 886-6159
U.S. EPA, Region 6	1445 Ross Ave.	6H-A	Dallas, TX 75202	(214) 655-6755
U.S. EPA, Region 7	726 Minnesota Ave.	RCRA Branch	Kansas City, KS 66101	(913) 551-7651
U.S. EPA, Region 8	999 18th St.	8-HWM-RI	Denver, CO 80202	(303) 293-1514
U.S. EPA, Region 9	75 Hawthorne St.	10th Floor, H21	San Francisco, CA 94103	(415) 744-2079
U.S. EPA, Region 10	1200 Sixth Ave.	WD-139	Seattle, WA 98101	(206) 553-0344

Table A-2. States/Territories covered by EPA region

EPA region	States/Territories
1	ME, VT, NH, MA, RI, CT
2	NY, NJ, U.S. Virgin Islands, Puerto Rico
3	PA, WV, VA, MD, DE, Washington DC
4	KY, TN, NC, SC, GA, AL, MS, FL
5	OH, IN, IL, MI, WI, MN
6	LA, AR, OK, TX, NM
7	NE, KS, MO, IA
8	ND, SD, MT, WY, CO, UT
9	AZ, CA, NV, HI, Guam, American Samoa
10	OR, ID, WA, AK

Additional Federal Offices

U.S. Department of Labor
Occupational Safety and Health Administration (OSHA)
200 Constitution Ave. NW
Washington, DC 20210
(202) 523-8036

U.S. Department of Transportation
U.S. Coast Guard
2100 Second St. SW
Washington, DC 20593
(202) 267-2200

Federal Environmental Hotlines

Acid Rain (General Information)	(617) 674-7377
Air Control Technology Assistance	(919) 541-0800
NPDES (Storm Water) Permitting	(703) 821-4823
RCRA (Hazardous Waste) Ombudsman	(800) 262-7937
Solid and Hazardous Waste (RCRA)	(800) 424-9346 / (703) 412-9810
Superfund (CERCLA)	(800) 424-9346 / (703) 412-9810
Underground Storage Tanks	(800) 424-9346 / (703) 412-9810
Wetland Information	(800) 832-7828

State and Territorial

Alabama Department of Environmental Management
1751 Cong. W. L. Dickenson Drive
Montgomery, AL 36130
(205) 271-7861

Alaska Department of Environmental Conservation
P.O. Box O
Juneau, AK 99811
(907) 465-2666

Arizona Department of Environmental Quality
2005 North Central Avenue
Phoenix, AZ 85004
(602) 257-2308

Arkansas Department of Pollution Control and Ecology
8001 National Drive
Little Rock, AR 72209
(501) 562-7444

California Air Resources Board
P.O. Box 2815
Sacramento, CA 95812
(916) 445-4383

California Water Resources Control Board
AST Program
P.O. Box 944212
Sacramento, CA 94244
(916) 227-4364

Colorado Department of Health
4210 East 11th Avenue
Denver, CO 80220
(303) 331-8500

Connecticut Department of Environmental Protection
165 Capitol Avenue
Hartford, CT 06106
(203) 566-2506

Delaware Department of National Resources and Environmental Control
89 Kings Highway
Dover, DE 19903
(302) 736-4791

District of Columbia Department of Consumer and Regulatory Affairs
5010 Overlook Avenue, SW
Washington, DC 20032
(202) 767-7370

Florida Department of Environmental Regulation
2600 Blair Stone Road
Tallahassee, FL 32399
(904) 488-4805

Georgia Department of Natural Resources
205 Butler Street, SE
Atlanta, GA 30334
(404) 656-6900

Hawaii State Department of Health
1270 Queen Emma Street
Honolulu, HI 96813
(808) 548-3676

Idaho Division of Environmental Quality
1410 North Hilton
Boise, ID 83706
(208) 334-5898

Illinois Environmental Protection Agency
2200 Churchill Road
Springfield, IL 62794
(217) 782-7326

Indiana Department of Environmental Management
105 South Meridian Street
Indianapolis, IN 46206
(317) 232-8384

Iowa Department of Natural Resources
900 East Grand
Des Moines, IA 50319
(515) 281-8935

Kansas Department of Health and Environment
Forbes Field, Building 740
Topeka, KS 66620
(913) 296-1593

Kentucky Department for Environmental Protection
316 St. Clair Mall
Frankfort, KY 40601
(502) 564-3382

Louisiana Department of Environmental Quality
P.O. Box 44096
Baton Rouge, LA 70804
(504) 342-1206

Maine Department of Environmental Protection
State House, Station 17
Augusta, ME 04333
(207) 289-2437

Maryland Department of the Environment
2500 Broening Highway
Baltimore, MD 21224
(301) 631-3255

Massachusetts Department of Environmental Protection
One Winter Street, 8th Floor
Boston, MA 02108
(617) 292-5630

Michigan Department of Natural Resources
P.O. Box 30028
Lansing, MI 48909
(517) 373-7023

Minnesota Pollution Control Agency
520 Lafayette Road North
St. Paul, MN 55155
(612) 296-7331

Mississippi Department of Environmental Quality
P.O. Box 10385
Jackson, MS 39289
(601) 961-5171

Missouri Department of Natural Resources
P.O. Box 176
Jefferson City, MO 65102
(314) 751-4817

Montana Department of Health and Environmental Sciences
Cogswell Building
Helena, MT 59620
(406) 444-3454

Nebraska Department of Environmental Control
301 Centennial Mall South
Lincoln, NE 68509
(402) 471-2189

Nevada Division of Environmental Protection
123 West Nye
Carson City, NV 89710
(702) 687-5065

New Hampshire Department of Environmental Services
Air Resources Division
64 North Main Street
Concord, NH 03302
(603) 271-1370

New Jersey Department of Environmental Protection
401 East State Street
Trenton, NJ 08625
(609) 292-5383

New Mexico Health and Environment Department
1190 St. Francis Drive
Sante Fe, NM 87503
(505) 827-0070

New York Department of Environmental Conservation
50 Wolf Road
Albany, NY 12233
(518) 457-7230

North Carolina Department of Environment, Health, and Natural Resources
P.O. Box 27687
Raleigh, NC 27611
(919) 733-3340

North Dakota Department of Health and Consolidated Laboratories
1200 Missouri Avenue
Bismark, ND 58502
(701) 224-2348

Ohio Environmental Protection Agency
1800 Watermark Drive
Columbus, OH 43226
(614) 644-2270

Oklahoma Department of Health
P.O. Box 53551
Oklahoma City, OK 73152
(405) 271-5220

Oregon Department of Environmental Quality
811 SW 6th Avenue
Portland, OR 97204
(503) 229-5388

Pennsylvania Department of Environmental Resources
P.O. Box 2357
Harrisburg, PA 17120
(717) 787-9702

Rhode Island Department of Environmental Management
291 Promenade Street
Providence, RI 02908
(401) 277-2808

South Carolina Department of Health and Environmental Control
2600 Bull Street
Columbia, SC 29201
(803) 734-4750

South Dakota Department of Water and Natural Resources
523 East Capitol Avenue
Pierre, SD 57501
(605) 773-3153

Tennessee Department of Health and Environment
701 Broadway
Nashville, TN 37247
(615) 741-3931

Texas Natural Resources Conservation Commission
Petroleum Storage Tank Division
P.O. Box 13087
Austin, TX 78711
(512) 239-2000

Utah Department of Health
P.O. Box 16690
Salt Lake City, UT 84116
(801) 538-6108

Vermont Agency of Natural Resources
103 South Main Street
Waterbury, VT 05676
(802) 244-8731

Virginia Department of Environmental Quality
Office of Spill Response and Remediation
P.O. Box 11143
Richmond, VA 23230
(804) 527-5314

Washington Department of Ecology
PV 11
Olympia, WA 98504
(206) 459-6256

West Virginia Department of Environmental Protection
Office of Air Quality
1558 Washington Street East
Charleston, WV 25311
(304) 558-4022

Wisconsin Department of Natural Resources
P.O. Box 7921
Madison, WI 53707
(608) 266-7718

Wyoming Department of Environmental Quality
122 West 25th Street
Cheyenne, WY 82002
(307) 777-7391

U.S. Virgin Islands Department of Planning and Natural Resources
Watergut Homes 1118
Christiansted, St. Croix
US VI 00820
(809) 773-0565

Environmental Regulatory Offices—Canada

Federal

Environment Canada
Oil, Gas and Energy Division
10th Floor, Place Vincent Massey
Ottawa, Ontario K1A 0H3
(819) 997-1221

Provinces and Territories

Alberta
Alberta Environment
5th Floor, Oxbridge Place
9820 - 106 Street
Edmonton, Alberta T5K 0G5
(403) 427-6182

Technical Services
705, 10808 - 99 Avenue
Edmonton, Alberta T5K 0G5
(403) 427-8265

British Columbia
B.C. Ministry of Environment
Mining and Chemical Industries
Environmental Protection Division
777 Broughton Street
Victoria, British Columbia V8V 1X5
(604) 387-9952

Ministry of Municipal Affairs, Recreation and Culture
Office of the Fire Commissioner
3rd Floor, 800 Johnson Street
Victoria, British Columbia V8V 1X4
(604) 356-9000

Manitoba
Department of Environment
Operations, Winnipeg Region
Building 2, 139 Tuxedo Avenue
Winnipeg, Manitoba R3N 0N6
(204) 945-7110

New Brunswick
Department of the Environment
Industrial Programs Section
364 Argyle Street
Fredericton, New Brunswick E3B 5H1
(506) 457-4848

Newfoundland
Department of Environment and Lands
Industrial Environmental Engineering Division
P.O. Box 8700
St. John's, Newfoundland A1B 4J6
(709) 729-2561

Northwest Territories
Government of Northwest Territories
Department of Safety and Public Services
Office of the Fire Marshal
Main Floor, Northway Building
Box 1302, 49th Avenue
Yellowknife, N.W.T. X1A 2L9
(403) 873-7472

Nova Scotia
Department of the Environment
Petroleum Storage Tank Systems
5th Floor, 5151 Terminal Road
Halifax, Nova Scotia B3J 3B7
(902) 424-5300

Ontario
Ministry of Consumer and Commercial Relations
Fuels Safety Branch
Mutual Group Centre, West Tower
4th Floor, 3300 Bloor Street West
Etobicoke, Ontario M8X 2X4
(416) 234-6030

Ministry of Environment
North Eastern Region
Sudbury Regional Office
199 Larch Street, 11th Floor
Sudbury, Ontario P3E 5P9
(705) 675-4501

Prince Edward Island
Department of the Environment
Environmental Protection Branch
P.O. Box 2000
Charlottetown, P.E.I. CIA 7N8
(902) 368-5057

Quebec
Department of Energy and Resources
Petroleum Products Branch
Room B405, 5700 - 4th Avenue West
Charlesbourg, Quebec G1H 6R1
(418) 643-3327

Saskatchewan
Department of Community Services
Fire Commissioner's Office
1870 Albert Street
Regina, Saskatchewan S4P 3V7
(306) 787-4516

Saskatchewan Environment and Public Safety
Chemical Management Section
Air and Land Protection Branch
3085 Albert Street
Regina, Saskatchewan S4S 0B1
(306) 787-6185

Yukon

Department of Community and Transportation Services
Public Safety Branch
Box 2703, 2nd Avenue
Whitehorse, Yukon Y1A 2C6
(403) 667-5824

APPENDIX **B**

AST Fabricating Companies, Field-Erected and Shop-Built

Tables B-1 through B-3 list member companies of the Steel Tank Institute, the Petroleum Equipment Institute, and the Steel Tank Association of Canada. Table B-4 is an alphabetical listing of major field-erected AST companies compiled from the 1993 ILTA Supplier Directory, the 1993 Thomas Register, various industry trade shows, and the authors' industry experience.

Table B-1. Steel Tank Institute member companies, United States

Company	Headquarters	Telephone	Number of locations	Shop-built ASTs	Field-erected ASTs
Ace Tank & Equipment	Seattle, WA	(206) 281-5000	3	yes	yes
Adamson Company, Inc.	Chester, VA	(804) 748-6453	3	yes	no
Alabama Tank	Atmore, AL	(205) 577-6415	1	yes	no
Anchorage Tank & Welding	Anchorage, AL	(907) 272-3543	1	yes	no
B & H Tank Corporation	Marion, IA	(319) 377-6357	1	yes	no
Bakersfield Tank Co.	Bakersfield, CA	(805) 589-9141	1	yes	no
Beta Tank, Inc.	Lebanon, PA	(717) 270-2691	1	yes	no
The Boss Group	Baltimore, MD	(800) 759-9988	1	yes	no
Bristol Tank & Welding	Langhorne, PA	(215) 752-9556	1	yes	no
Brothers Manufacturing	Hermansville, MI	(906) 498-7771	1	yes	yes
Brown-Minneapolis Tank	St. Paul, MN	(612) 454-6750	4	yes	yes
Certified Equipment	Springfield, IL	(217) 525-1433	1	yes	no
Chicago Boiler	Lincolnshire, IL	(708) 913-0100	1	yes	no
Clawson Tank Company	Clarkston, MI	(313) 625-8700	1	yes	no
Derson Tank	Watertown, MN	(612) 955-2668	1	yes	no
Drummond Weldsteel Inc.	Lebanon, NH	(603) 448-1440	1	yes	yes
Eaton Metal Products Co.	Denver, CO	(303) 296-4800	5	yes	yes
Eidson Steel Products	Albuquerque, NM	(505) 873-0160	1	yes	yes
Empire Steel Manufacturing	Billings, MT	(406) 252-0101	1	yes	no
Fargo Tank Company	Fargo, ND	(701) 282-2345	1	yes	no
Fedco Manufacturing Corp.	Somersworth, NH	(603) 692-2921	1	yes	yes
General Industries	Goldsboro, NC	(919) 735-2882	1	yes	no
Gordon Metal Fabricators	Spokane, WA	(509) 244-4556	1	yes	yes
Greer Tank, Inc.	Fairbanks, AK	(907) 243-2455	2	yes	no
Hall Tank Company	N. Little Rock, AR	(501) 945-3211	1	yes	no
Hamilton Tanks	Columbus, OH	(614) 445-8446	1	yes	no
Highland Tank & Mfg. Co.	Stoystown, PA	(814) 893-5701	6	yes	no
J. L. Houston Company	Hopkins, MO	(816) 778-3393	1	yes	no
Huron Culvert & Tank Co.	Huron, SD	(605) 352-8643	1	yes	no
International Product Specialists (IPS)	Honey Creek, WI	(414) 534-3130	1	yes	no
Imperial Tank Co.	San Antonio, TX	(512) 661-2316	1	yes	no
Kennedy Tank & Mfg. Co.	Indianapolis, IN	(317) 787-1311	1	yes	yes
Kleespie Tank & Petroleum	Morris, MN	(612) 589-2206	1	yes	no
Kohlhass Corporation	Albuquerque, NM	(505) 243-3766	2	yes	no
Lancaster Tanks & Steel	Lancaster, NY	(716) 683-2033	1	yes	no
Lannon Tank Corporation	Lannon, WI	(414) 251-7890	1	yes	no
Massachusetts Engineering	Avon, MA	(508) 580-0550	1	yes	no
Metal Products Company	Suwanee, GA	(404) 945-8383	1	yes	no
Mid-South Steel Products	Cape Girardeau, MO	(314) 335-5529	1	yes	no
Modern Welding Company	Owensboro, KY	(502) 685-4400	7	yes	yes
Mohawk Metal Products	Utica, NY	(315) 793-3000	1	yes	no
Mosier Brothers	Woodlake, CA	(209) 564-3304	1	yes	no
Newberry Tanks & Equip.	West Memphis, AR	(501) 735-4473	1	yes	no

Table B-1. *(continued)*

Company	Headquarters	Telephone	Number of locations	Shop-built ASTs	Field-erected ASTs
O'Day Equipment, Inc.	Fargo, ND	(701) 282-9260	4	yes	no
Palmer Mfg. & Tank	Garden City, KS	(316) 275-7461	1	yes	yes
Pee Dee Tank Company	Florence, SC	(803) 669-3381	1	yes	no
Perkins Welding Works	Sacramento, CA	(916) 383-5413	1	yes	no
George I. Reitz & Sons, Inc.	Brookville, PA	(814) 849-2308	1	yes	no
Reliable Steel Fabricators	Olympia, WA	(206) 352-7575	1	yes	yes
Service Welding, Inc.	Louisville, KY	(502) 585-4295	1	yes	no
Smith Tank Company	Tulsa, OK	(918) 446-4484	1	yes	no
Southern Tank & Mfg. Inc.	Owensboro, KY	(502) 684-2321	1	yes	no
Specialty Tank & Equip. Co.	Jacksonville, FL	(904) 353-0108	1	yes	no
Stanwade Metal Products	Hartford, OH	(216) 772-2421	1	yes	no
Starco, Inc.	Rhome, TX	(817) 636-2215	1	yes	no
Steel Tank & Fabricating Co., Inc.	Mancelona, MI	(616) 587-9191	1	yes	no
Steel Tank & Fabricating Corp.	Columbia City, IN	(219) 248-8971	1	yes	yes
Topaz Tank & Mfg.	Boise, ID	(208) 362-1738	1	yes	no
Usemco, Inc.	Tomah, WI	(608) 372-5911	1	yes	no
Watco Tanks, Inc.	LaVernia, TX	(210) 779-2411	1	yes	no
We-Mac Manufacturing Co.	N. Kansas City, MO	(816) 221-1850	3	yes	no

Source: Steel Tank Institute Membership Roster, 1993.

Table B-2. Petroleum Equipment Institute manufacturing members of ASTs (not listed under STI)

Company	Headquarters	Telephone	Number of locations	Shop-built ASTs	Field-erected ASTs
Agape Manufacturing Co.	Poway, CA	(619) 748-8815	1	yes	no
Arco-Power	Smithtown, NY	(516) 366-4362	1	yes	no
Bendel Corporation	Charlotte, NC	(704) 596-5112	1	yes	no
B.R.E. Products, Inc.	Huntington Beach, CA	(714) 965-7179	1	yes	no
Convault, Inc.	Denair, CA	(209) 632-7571	1*	yes	no
Ecovault (Div. of Unisil)	Reston, VA	(703) 264-1200	1*	yes	no
Hallmark Industries	Gardena, CA	(800) 872-2270	1*	yes	no
Industrial Environmental Supply	Greensboro, NC	(910) 274-4817	1	yes	no
Joor Manufacturing, Inc.	Escondido, CA	(619) 745-0971	1	yes	no
LRS, Inc.	S. El Monte, CA	(800) 966-4577	1	yes	no
Hoover Containment Systems	Severna Park, MD	(410) 987-3800	1	yes	no
Trusco Tank, Inc.	Fresno, CA	(209) 264-4741	2	yes	yes
Zorn Industries	Elkhart, IN	(219) 266-1433	1	yes	no

Source: Petroleum Equipment Institute Directory, 1993.

*These companies license their products to other manufacturers at locations around the country.

Table B-3. Steel Tank Association of Canada member companies

Company	Headquarters	Telephone	Number of locations	Shop-built ASTs	Field-erected ASTs
MFM Industries	Montcon, NB	(506) 857-2700	1	yes	yes
Steel & Engine Production Ltd.	Liverpool, NS	(902) 354-3483	1	yes	no
Clemmer Industries	Waterloo, Ont.	(519) 884-4320	3	yes	yes
Drummond Welding	Longueuil, Que.	(514) 526-4411	1	yes	yes
DTE Industries Ltd.	Scarborough, Ont.	(416) 757-6278	1	yes	no
Granby Steel Tanks Inc.	Granby, Que.	(514) 378-2334	1	yes	no
Hassan Steel Fabricators	London, Ont.	(515) 451-3100	1	yes	no
Soudure F.M. Welding Inc.	Montreal, Que.	(514) 648-4987	1	yes	no
Tidy Steel Fabricating/Regal Tanks	Surrey, BC	(604) 580-9733	1	yes	yes
Westeel	Calgary, Alb.	(403) 278-4500	2	yes	yes

Source: Steel Tank Association of Canada, 1993.

Table B-4. Major field-erected AST companies, United States

AST company	Headquarters address	City/state/zip code	Telephone
Advance Tank & Construction Co.	P.O. Box 219	Willmington, CO 80549	(303)568-3444
American Tank & Vessel, Inc.	P.O. Box 910	Mobile, AL 36601	(205)432-8265
Baker Tank Company/Altech	P.O. Box 40	Arp, TX 75750	(903)859-2111
Brown-Minneapolis Tank Co. (BMT)	P.O. Box 64670	Saint Paul, MN 55164	(612)454-6750
Caldwell Tanks, Inc.	400 Tower Rd.	Louisville, KY 40219	(502)964-3361
Challenger Tank & Mfg. Company	P.O. Box 249	Whitehouse, TX 75791	(903)839-7291
Chicago Bridge & Iron Co. (CBI)	800 Jorie Blvd.	Oak Brook, IL 60522	(312)572-7000
Consolidated Fabricators & Constructors, Inc.	640 S. Wheeler Place	Griffith, IN 46319	(219)924-2884
Continental/Freemont, Inc.	P.O. Box 827	Metter, GA 30439	(419)448-4045
Diamond Tank & Vessel, Inc.	4533 Pasadena Blvd.	Pasadena, TX 77503	(713)479-4010
Erected Steel Products of Alabama Inc.	P.O. Box 360347	Birmingham, AL 35236	(205)945-7494
Fisher Tank Company	3131 W. Fourth St.	Chester, PA 19013	(215)494-7200
Graver Tank & Mfg. Co., Inc.	P.O. Box 1764	Houston, TX 77251	(713)474-5121
HMT, Inc.	4422 FM 1960 W., #350	Houston, TX 77068	(713)893-9150
International Tank Services, Inc.	P.O. Box 5065	Lima, OH 45802	(419)223-8251
Jersey Tank Fabricators Inc.	P.O. Box 64	Bayonne, NJ 07002	(908)561-2865
Matrix Service Company	10701 E. Ute St.	Tulsa, OK 74116	(918)838-8822
Nooter Corporation	P.O. Box 451	Saint Louis, MO 63166	(314)621-6000
Pasadena Tank Corporation	8123 Hillsboro	Houston, TX 77029	(713)675-4096
Pawnee Tank Company	554 Illinois St.	Pawnee, OK 74058	(918)762-2182
Pitt-Des Moines, Inc. (PDM)	3400 Grand Ave.	Pittsburgh, PA 15225	(412)332-3000
Pittsburg Tank & Tower Company	P.O. Box 913	Henderson, KY 42420	(502)826-9000
Tampa Tank & Welding, Inc.	P.O. Box 5062	Tampa, FL 33675	(813)623-2675
Tank Builders, Inc.	P.O. Box 1527	Euless, TX 76039	(817)571-4044
Witherup Fabrication & Erection, Inc.	P.O. Box 55	Kennerdell, PA 16374	(814)385-6601

*These companies license their products to other manufacturers at locations around the country.

APPENDIX C

AST Associations

The following associations impact the AST industry:

American National Standards Institute (ANSI)
1430 Broadway
New York, NY 10018
(212) 642-4900

American Petroleum Institute (API)
1220 L St. NW
Washington, DC 20005
(202) 682-8000

American Society of Civil Engineers (ASCE)
345 East 47th Street
New York, NY 10017
(212) 705-7071

American Society of Mechanical Engineers (ASME)
345 East 47th Street
New York, NY 10017
(212) 705-7722

American Society for Testing and Materials (ASTM)
1916 Race Street
Philadelphia, PA 19103
(215) 299-5400

Building Officials and Code Administrators International (BOCA)
4051 West Flossmoor Road
Country Club Hills, IL 60477
(708) 799-2300

Chemical Manufacturers Association (CMA)
2501 M Street NW
Washington, DC 20037
(202) 887-1108

Environmental Defense Fund
1875 Connecticut Avenue NW, Suite 1016
Washington, DC 20009
(202) 387-3500

Independent Liquid Terminals Association (ILTA)
133 15th Street NW, Suite 204
Washington, DC 20005
(202) 659-2301

Independent Refiner Marketers Association (IRMA)
9901 Interstate Highway 10 West, Suite 800
San Antonio, TX 78230
(512) 558-2860

International Fire Code Institute (IFCI)
5360 South Workman Mill Road
Whittier, CA 90601
(310) 699-0124

National Association of Corrosion Engineers (NACE)
P.O. Box 218340
Houston, TX 77218
(713) 492-0535

National Fire Protection Association (NFPA)
Batterymarch Park
P.O. Box 9101
Quincy, MA 02269
(617) 377-3000

National Leak Detection Association
7685 Fields Ertel Road
Cincinnati, OH 45241
(513) 489-9844

National Petroleum Council
1625 K Street NW, Sixth Floor
Washington, DC 20006
(202) 393-4100

Petroleum Equipment Institute (PEI)
P.O. Box 2380
Tulsa, OK 74101
(918) 494-9696

Petroleum Marketers Association of America (PMAA)
1120 Vermont Avenue NW, Suite 1130
Washington, DC 20005
(202) 331-1198

Society of Independent Gasoline Marketers of America (SIGMA)
11911 Freedom Drive, Suite 590
Reston, VA 22090
(703) 709-7000

Society of Petroleum Engineers
P.O. Box 833836
Richardson, TX 75083
(214) 669-3377

Southern Building Code Congress International
900 Montclair Road
Birmingham, AL 35213
(205) 591-1853

Steel Tank Institute (STI)
570 Oakwood Road
Lake Zurich, IL 60047
(708) 438-8265

Underwriters Laboratories (UL)
333 Pfingsten Road
Northbrook, IL 60062
(708) 272-8800

APPENDIX D

Canadian Federal AST Legislation and AST Associations

This appendix is designed to explain the AST regulatory market in Canada and was provided by Mr. Kelly Karr, P.E., who is the head of the Downstream Petroleum Section at Environment Canada in Ottawa. Mr. Karr is also the cochair of the CCME National Task Force on Storage Tanks and chair of the federal government's Federal Facilities Storage Tank Committee. Following the Canadian AST codes discussion, a Canadian associations directory is provided.

Canadian AST Codes and Regulations

In Canada, the regulation of storage tanks is primarily a provincial responsibility. However, the federal government plays an important role in the overall management of aboveground storage tanks by coordinating the development of two national model codes. One of these codes in the *National Fire Code of Canada* (NFC). The other code is the *Environmental Code of Practice for Aboveground Storage Tank Systems Containing Petroleum Products* (ECP), which was published in 1994 by the Canadian Council of Ministers of the Environment (CCME).

Both model codes are written in a form suitable for adoption by the applicable provincial and territorial regulatory authority.

The National Fire Code of Canada

The *National Fire Code of Canada* is managed by the Canadian Codes Centre of the National Research Council of Canada. The NFC is presently reviewed and amended on a continuous five-year cycle.

The *National Fire Code of Canada,* part 4, addresses the storage of flammable and combustible liquids. The NFC is similar to the U.S. NFPA 30 Standard. The NFC addresses ASTs from a fire and safety perspective at a new installation. It is left entirely to the provincial regulatory authority to determine whether or not new or more stringent amendments to the NFC should be applied to existing storage tank systems. The NFC is adopted (completely or with minor amendments) by most provinces via a provincial fire code.

A Canadian Treasury Board Directive requires tank owners on federal lands to follow the National Fire Code.

CCME Environmental Code of Practice

CCME's *Environmental Code of Practice for Aboveground Storage Tank Systems,* published in 1994, is a model set of technical requirements designed to protect the environment by preventing product releases from AST systems. A similar code of practice for underground storage tank systems was published by CCME in 1988 and then revised in 1989 and 1993.

The Canadian Council of Ministers of the Environment (CCME) is the major intergovernmental forum in Canada for discussion and joint action on environmental issues of national, international, and global concern. Environment ministers from each of the ten provinces, the two territories, and the federal government participate in CCME meetings at least twice a year. They discuss environmental issues, exchange information, make decisions, and establish policies for work to be carried out under the auspices of CCME. The presidency and other official posts of CCME are rotated annually among member governments.

The National Task Force on Storage Tanks reports to the Environmental Protection Committee, one of two permanent steering committees. The National Task Force has members from virtually all stakeholder groups. The provincial governments typically have representatives from the Ministry of the Environment and the ministry responsible for fire safety and/or fuel safety. There are also members from major oil companies, industry associations (the Canadian Petroleum Products Institute), the Canadian Codes Centre (NFC), the Steel Tank Association of Canada (STAC), and the Canadian standards writing organizations (Underwriters Laboratories of Canada and the Canadian Standards Association), as well as representatives from federal government departments that are responsible for large numbers of storage tanks (i.e., Transport Canada and National Defence). Environment Canada cochairs and acts as the secretariat for the National Task Force.

A CCME Environmental Code of Practice (ECP) comes into effect when the provincial regulatory authority adopts the code by reference. However, the provincial regulatory authorities usually take the recommendations from the ECP and rewrite them into their provincial storage tank regulations. A Canadian Environmental Protection Act guideline has brought the ECP into effect for all aboveground storage tank systems installed on federal lands.

Some of the key provisions of the 1994 CCME ECP for AST systems are as follows (*note*: 1 U.S. gallon = 3.78 liters):

- The code is intended for atmospheric aboveground storage tanks having a capacity greater than 230 liters (61 gallons) that are designed to be installed at a fixed location outdoors. However, storage tanks at a refinery or a process plant are not included.
- Registration is required for new and existing aboveground storage tanks having a capacity greater than 4000 liters (1058 gallons).
- Storage tanks having a capacity greater than 4000 liters (1058 gallons) shall control emissions of volatile organic compounds (VOC) in conformance with another CCME code of practice, *Guideline for Controlling Emissions of Volatile Organic Compounds from Aboveground Storage Tanks*. Table D-1 illustrates the requirements of this code.
- Storage tanks having a capacity greater than 4000 liters (1058 gallons) shall have secondary containment with leak detection.

- Field-erected ASTs with pipeline delivery shall have automatic alarms.
- A visual and audible overfill alarm is required for other tanks that have a capacity greater than 50,000 liters (13,228 gallons).
- Tank bottom water shall be segregated from rainwater and disposed of according to provincial regulations.
- Reconstructed, relocated, or refurbished storage tanks shall be in conformance with the applicable ULC standard or API 653.
- As-built drawings shall be provided to the regulatory authority.
- Underground piping 75 mm diameter and smaller connected to an aboveground tank system shall have secondary containment with leak detection.
- Underground piping larger than 75 mm diameter shall have either:
  (a) Secondary containment with leak detection, or
  (b) Cathodic protection with periodic leak detection or testing.
- The code contains details on what is acceptable secondary containment and leak detection for ASTs and connected underground piping.
- Existing field-erected ASTs must be upgraded to new standards within 15 years from the date when the authority adopts the code.
- Existing shop-fabricated ASTs must be upgraded to new standards within ten years from the date when the authority adopts the code.
- Overfill protection must be installed on tank systems within five years from the date when the authority adopts the code.
- Periodic internal and external recordkeeping is required.
- Field-erected tanks shall be inspected in accordance with API 653.
- Oil–water separators shall be designed to discharge an effluent having less than 15 ppm oil and grease.
- The code requires the periodic removal of the oil layer and the sludge from an oil–water separator (at least every two years).
- Records must be kept of tank inspections, maintenance, and the removal of the oil layer and sludge from an oil–water separator.
- Spills, leaks, and overfills of more than 100 liters (26 gallons) shall be reported to the regulatory authority.
- Field-erected tanks that are out of service for more than one year shall pass an internal inspection in conformance with API 653.
- When a storage tank system is permanently withdrawn from service, the tank, piping, and product transfer equipment all shall be removed.

Canadian AST Associations

For a listing of the federal and provincial environmental offices (i.e., Environment Canada), refer to Appendix A. The following associations impact the Canadian AST industry:

Canadian Council of Ministers of the Environment (CCME)
326 Broadway, Suite 400
Winnipeg, Manitoba R3C 0S5
(204) 948-2090

Canadian General Standards Board (CGSB)
1402 - 222 Queen Street
Ottawa, Ontario K1A 1G6
(613) 941-8640

Canadian Petroleum Products Institute (CPPI)
1000 - 275 Slater Street
Ottawa, Ontario K1P 5H9
(613) 232-3709

Canadian Standards Association (CSA)
178 Rexdale Boulevard
Rexdale, Ontario M9W 1R3
(416) 747-4363

National Fire Code of Canada (NFC)
c/o National Research Council of Canada
Ottawa, Ontario K1A 0R6
(613) 993-2463

Underwriters Laboratories of Canada (ULC)
7 Course Road
Scarborough, Ontario M1R 3A9
(416) 757-3611

Table D-1. Summary of CCME Guideline for Controlling Emissions of VOC's from ASTs

Areas of application	This guideline is applicable to new tanks everywhere in Canada, and to existing tanks in the Lower Fraser Valley, the Windsor–Quebec City corridor, and the Saint John area.
Timing	• New tanks: one year after the publication of this guideline by the CCME. • Existing floating roof tanks: when seals are worn, or by December 31, 2004, whichever is later. • Existing vapor control/balancing systems: by December 31, 2004. • Existing fixed roof tanks: when degassed for scheduled maintenance, or by December 31, 2004, whichever is earlier. • Tanks subjected to the guideline to comply with the inspection and recordkeeping requirements one year after the publication of this guideline by CCME.
Submerged fill pipe	All tanks with capacities greater than 4 m^3 storing a liquid with vapor pressure ≥ 10 kPa shall have a submerged fill pipe.
Pressure–vacuum vent	• Vertical tanks with capacities greater than 4 m^3 but diameters less than 4 m storing a liquid with vapor presure ≥ 10 kPa shall have a pressure–vacuum vent. • Nonvertical tanks with capacities ≥ 4 m^3 but < 75 m^3 storing a liquid with vapor pressure ≥ 10 kPa shall have a pressure–vacuum vent.
Recordkeeping	• Tank/system construction and modification date for the life of the tank/system. • Inspection data for the last ten years. • Tank throughput by liquid type and vapor pressure for the last three years.

Technology:	**Internal floating roof**	**External floating roof**	**Vapor control system**	**Vapor balancing system**
Requirements for vertical tanks: diameter ≥ 4.0 but <9.0 m, vapor pressure ≥ 10 kPa but < 76 kPa	yes	yes	yes	yes
Requirements for vertical tanks: diameter ≥ 9.0 m, vapor pressure ≥ 10 kPa but < 76 kPa	yes	yes	yes	no
Requirements for nonvertical tanks: capacity ≥ 75 but < 500 m^3, vapor pressure ≥ 10 kPa but < 76 kPa	N/A	N/A	yes	yes
Requirements for nonvertical tanks: capacity ≥ 500 m^3, vapor pressure ≥ 10 kPa but < 76 kPa	N/A	N/A	yes	no
Requirements for all tanks: capacity ≥ 75 m^3, vapor pressure ≥ 76 kPa	no	no	yes	no
Specifications	Primary seal, secondary seal (with vapor primary), gaps, covers	Primary seal (liquid or shoe), secondary seal, gaps, covers	Vapor control efficiency ≥ 95% by weight	Vapor balancing efficiency ≥ 90% by weight for the working losses
Inspections	Once every ten years and visual annually, or LEL measurements	Visual annually, primary seal once every five years, and secondary seal annually	Annually	Annually

Source: 1994 Draft of CCME Guideline; approval by National Air Issues Coordinating Committee still required.

Note: This table is provided as an overview for the convenience of the reader. Refer to the guideline text for explicit requirements.

APPENDIX E

National Fire Code Comparison Chart by the Steel Tank Institute

The following chart has been reprinted with permission from the Steel Tank Institute and its publication *Tank Talk*. The chart focuses on code requirements for aboveground storage tanks dispensing fuels. This chart was developed to help AST owners understand the various fire codes and how they are related.

Code Requirements on Aboveground Storage Tanks Dispensing Fuels

	NFPA 30A 1993 EDITION	BOCA 1993 EDITION	SBCCI/SFPC 1994 EDITION	UFC 1994 EDITION
APPROVAL/GENERAL PROVISIONS	Tanks located aboveground at automotive service stations must have approval of the authority having jurisdiction (AHJ) and installed in accordance with Sections 2-4. (5)(6)	Aboveground tanks shall be permitted for the storage of motor fuels on premises to which the public does not have access when installed in a special enclosure constructed in accordance with NFiPA 30A (1990 edition) listed in Chapter 44 or protected by a listed and approved tank enclosure assembly providing fire resistance protection of not less than 2 hours from exposure to a flammable liquid pool fire provided that specific approval is obtained.	Protected tanks must be listed OR approved. UL 2085 is not referenced. Special enclosures must meet the construction requirements of 907.3.3. Assumption that tank must meet UL 142 per 907.3.4 (See Paragraph 102)	The dispensing of Class I and Class II liquids into motor vehicles from aboveground tanks is prohibited, except dispensing from special enclosures is allowed when underground tanks are impractical. However, Appendix II-F, which establishes requirements for dispensing from protected tanks is provided if local agencies wish to allow such dispensing by local ordinance.
TERMINOLOGY	**Fire Resistant Tank (Protected Tank)** Fire Resistant Tank that provides the required fire resistive protection, shall prevent release of liquid, failure of the primary tank, failure of the supporting structure, and impairment of venting for 2 hours using the fire exposure environment described in UL 2085 or equivalent. **Vault** NFPA 30A has elaborate language defining requirements. The vault must completely enclose each tank, either above or below grade, with at least 6" of concrete. The vault must have continuous ventilation, secured against unauthorized entry, and be equipped with a liquid detection system.	References NFPA 30A, 1990 edition, which does not incorporate any terminology for fire resistant tanks. **Special Enclosure** Constructed per NFPA 30A **Approved Tank Enclosure** Assembly fire-rated for 2 or more hours from exposure to a flammable pool fire.	**Fire Resistant Tank (Protected Tank)** Thermally protected aboveground storage tank (AST) listed or approved for aboveground use. "Approval" here means that the authority having jurisdiction (AHJ) can approve a non-listed tank. Thermally protected materials shall be weather resistant and resist impact of hose streams. **Special Enclosure** As described per section 907.3.5.1 for installation inside buildings.	**Protected Tank (1)** Tank system consisting of a primary tank protected from physical damage and from high intensity liquid pool fire exposure. The system must be listed and meet requirements of UFC Standard A-79-7. "Test Requirements for Protected Motor Vehicle Fuel Storage Tanks." 2-hour fire exposure, average thermocouple not to exceed 260°F with 400°F maximum. Structural integrity maintained and emergency venting operational. Post fire hose stream and leak test. **Special Enclosure** Tank enclosed by 6" of liquid tight, vapor tight concrete.
INSTALLATION	Per NFPA 30 (references PEI RP200-92, "*Recommended Practices for Installation of Aboveground Storage Systems for Motor Vehicle Fueling*")	All fuel-dispensing systems shall be installed in accordance with the mechanical code and NFiPA 30 or 30A listed in Chapter 44.	Follow manufacturer's instructions.	Fuel dispensing systems are to be installed in accordance with Article 79 which incorporates all the motor vehicle service station controls. However, tank installation only per Appendix requirements.
MAXIMUM CAPACITIES	12,000 gal individual/ public & private (5)(6) 40,000 gal aggregate	6,000 gal individual/ private 18,000 gal aggregate	6,000 gal individual/ private 18,000 gal aggregate	10,000 gal individual/ public and private 40,000 gal aggregate (2)

Code Requirements on Aboveground Storage Tanks Dispensing Fuels

	NFPA 30A 1993 EDITION	BOCA 1993 EDITION	SBCCI/SFPC 1994 EDITION	UFC 1994 EDITION
SPILL PREVENTION	Containment/spill control per NFPA 30 Overfill—alarm at 90% capacity & shut-off at 98%	Tanks shall be provided with automatic fuel shut-off devices capable of stopping the delivery of fuel into the storage tank at a 90% tank capacity.	No mandatory containment/spill control Overfill—shut-off at 90% capacity, no alarm	Containment—5 gal capacity or greater Overfill—alarm at 85% capacity & shut-off at 90% (3)
SEPARATION DISTANCE	**Uninsulated Tanks/ Private & Public (UL 142) (5):** • 100′ from property line • 50′ from buildings on same property • 50′ from fuel dispensers • 50′ from nearest side of public way **Fire Resistant Tanks/ Public (Protected Tank):** • 50′ from property line • 25′ from buildings on same property (4) • 25′ from fuel dispensers **Tanks in Vaults/Public** (5) **(Uninsulated):** • 50′ from property line • 25′ from buildings on same property (4) • 25′ from fuel dispensers **Fire Resistant Tanks and Tanks in Vault/Private:** • Same as with public facilities, except dispenser can be mounted directly atop tank/vault	**Special Enclosure/Private** • 5′ from building • 5′ from property line • 3′ between tank assemblies • 100′ between 18,000 gallon aggregates • Dispensing devices can be mounted directly atop tank **ASTs at Retail Facilities** • Not permitted	**Insulated Tanks/Private (Protected Tank)** • 5′ from building property line • 100′ between 18,000 gallon aggregates • 3′ between each tank assembly • 0′ for dispensing devices ***Special Enclosures*** For inside buildings only **Tanks in Vault/Private** • Subject to proof of 2-hour fire resistance **ASTs at Retail Facilities** • Not Permitted	**Protected Tank** < 6,000 gallons • 15′ from property line • 5′ from building or public way • 3′ between each tank > 6,000 gallons • 50′ from property line • 25′ from building or public way • 3′ between each tank • 100′ between 40,000 gallon aggregates • Dispensing devices can be mounted directly atop tank
PIPING	• Openings in tank top only • Expansion relief devices and anti-syphon devices required • Route piping to minimize exposure from damage • Listed emergency shut-off valves at dispensers where submersible pumps are used	• Openings in tank top only • Anti-syphon devices required • Additional piping requirements per NFPA 30A, 1990 edition. For instance, piping shall be located so as to be protected from damage.	• Openings in tank top only • Anti-syphon devices required	• Openings in tank top only • Anti-syphon devices required • Guard posts for connected piping subject to vehicle impact • Installed in accordance with Article 79 which addresses construction, corrosion, supports, joints, etc.
PHYSICAL PROTECTION	6′ high security fence located at least 10′ from tank and a gate that is properly secured against unauthorized entry. In addition, provide protection against vehicular collision by suitable barriers.	No security fence required. Guard posts 3′ from the tank and minimum 30′ in height to withstand 12,000 pound force.	No security fence required. Guard posts within 3′ from tank spaced 4′ on centers and minimum 30′ in height to withstand 12,000 pound force.	No security fence required. Guard posts within 5′ of tank, spaced no more than 4′ on center, not less than 3′ in height, constructed of not less than 4′ diameter steel and concrete filled.
TANK FILLING OPERATIONS	Separation of delivery vehicle by 25′ except gravity feed with underground (UG) vaults enables 0′ separation from tank. Liquid tight connection options.	No minimum separation distance of delivery vehicles from tanks specified. Transfer via fixed pumps designed to prevent leakage.	No minimum separation of delivery vehicle from tank. Transfer via fixed pumps designed to prevent leakage.	Separation of 25′ from parked tank vehicles except when the tank is being filled from the vehicle. Fill pipe shall be provided with a means of making a direct connection.

Code Requirements on Aboveground Storage Tanks Dispensing Fuels

	NFPA 30A 1993 EDITION	BOCA 1993 EDITION	SBCCI/SFPC 1994 EDITION	UFC 1994 EDITION
MISCELLANEOUS	Requirements for fuel dispensing devices, electrical equipment, vapor recovery and processing systems, and operational requirements.	Requirements for fuel dispensing devices, electrical equipment, vapor recovery and processing systems, and operational requirements.	Requires plans to be submitted for permit application incorporating details on vehicle access, seismic design, secondary containment, venting and vapor recovery and emergency controls required by fire official. It appears that fire officials would use their judgement to assess whether these requirements are adequate as no guidance language is given. Existing language for dispensing and controls remains.	Requirements for fuel dispensing, electrical equipment, vapor recovery, operation, labeling, signage and permits.

(1) Provisions are currently being considered for adoption for vaults, for the next full edition of the Uniform Fire Code (UFC).
(2) Special enclosure maximum capacities are 6,000/18,000 gallons, respectively.
(3) A code change is being processed which, if adopted, will allow for reduced flow at 85 percent and an alarm at 90 percent capacity for the next edition of the UFC.
(4) Includes 25 feet from nearest side of a public way.
(5) Section 9-3.5 enables tanks up to 6,000 gallon/12,000 aggregate to be installed at private fleet facilities and separated per Tables 2-6 in NFPA 30.
(6) When installation of tanks is impractical because of property or building limitations, tanks may be installed in buildings when enclosed within 6" of concrete (special enclosures) which is liquid and vapor tight. Tanks in parking facilities of large buildings storing Class I liquids are limited to 6,000 gallon/18,000 aggregate.

APPENDIX F

Petroleum Equipment Institute Dike Sizing Calculations

This appendix can be helpful in determining the appropriate size of a dike containment area for both vertical and horizontal ASTs. Usually, AST manufacturers will not offer this service. The following has been reprinted with permission from the Petroleum Equipment Institute.

A.1 Volume Calculation. The purpose of the volume calculation is to determine the adequacy of the diked area enclosing more than one tank. Where the diked area encloses more than one tank, the volume of the tanks, *other than the largest tank*, below the top of the dike, is deducted from the total capacity available to contain a release since it is unavailable to contain a release.

A.2 Information Available. As a guide to the procedure by which dike-size calculation is to be made, follow this example. Three tanks are to be contained in a dike. One is a 10,000 gallon capacity tank, nine feet in diameter and 21 feet long. Two are 6,000 gallon capacity tanks, eight feet in diameter and 16 feet long. To allow visual inspection of all tank surfaces, and to provide space for piping and other equipment, tanks are to be spaced three feet apart and three feet from the toe of the dike. Calculations are given for both vertical and horizontal tanks.

A.3 Required Containment Volume. The minimum containment volume required is equal to the volume which can be released into the diked area from the largest tank. Additional capacity may be required for the retention of rainwater or to provide an additional margin of safety. In this illustration, the additional required capacity has been established at ten percent.

Formula: Volume of a cylinder $V = \pi r^2 h$

where:

π = 3.14

r = Radius in feet

h = Height in feet

Therefore:

$$V = 3.14\ (4.5)^2 \times 21.00 = 1{,}335.3 \text{ cubic feet}$$

In this case, a manufacturer's tank chart was available which indicated that the actual capacity is 10,033 gallons. Using a conversion factor of 7.48 gallons per cubic foot, the calculated volume is 1,341.3 cubic feet. We have used the more conservative of the two values for our calculations. We also added ten percent to the diked capacity to accommodate rainwater. We determine that the containment required (V_c) is 1,487.4 cubic feet.

$$V_c = 1{,}341.3 + 10\% = 1{,}475.4 \text{ cubic feet}$$

A.4 Vertical Tanks. The shape of the diked area will depend upon what fits best on the particular site. The rectangular configuration was selected for our example.

Formula: Rectangular area = $A = L \times W$

where:

A = Area in square feet inside the tank

L = Length in feet = 37′0″

W = Width in feet = 15′0″

Gross containment area occupied = 37′0″ × 15′0″, or 555.0 square feet

A.5 Unavailable Volume. The volume occupied by tanks, *other than the volume of the largest tank* (V_u), is not available to contain a release and must be deducted from the volume of the diked area. In our example, the two 6,000 gallon capacity tanks are eight feet in diameter.

Formula: Area of the base of a cylinder = $A = \pi r^2$
Therefore:

$$V_u = 3.14\ (4.00)^2 \times 2 \text{ tanks} = 100.5 \text{ square feet}$$

The net diked area available for containment is 555.0 minus 100.5, or 454.5 square feet.

A.6 Calculation of Dike Height. The required dike height can be calculated by dividing the volume of the largest tank plus ten percent by the net containment area available, where height of dike wall = H:

$$H = \frac{V_c}{A - V_u} = \frac{1{,}475.4}{555.0 - 100.5} = 3.25 \text{ feet, or } 3'3''$$

Dimensions for containment installation:
Length = 37′0″
Width = 15′0″
Height = At least 3′3″

A.7 Horizontal Tanks. Calculations used to determine the capacity of the diked area and height of dikes, where horizontal tanks are to be installed, are similar to the calculations for vertical tanks. Again it is necessary to determine the capacity of the tanks, *other than the largest tank,* below the height of the dike. The simplest way to do this, using the manufacturer's tank conversion chart, is to convert the height from inches to gallons, and then to cubic feet, using a conversion factor of 7.48 gallons per cubic foot. The volume of a dike for horizontal tanks is computed from the same information provided.

A.8 Horizontal Tank Dike Dimensions. The required containment volume remains the same (1,475.4 cubic feet) since the volume of the largest tank and ten percent allowance are unchanged. The tanks in this example are installed two inches above the dike floor, and we selected the layout that best suits the property. For this initial calculation, disregard the volume of the other tanks below the top of the dike.

Formula: Rectangular area = $A = L \times W$

where:

A = Area in square feet inside the tank
L = Length in feet = 37′0″
W = Width in feet = 27′0″

Therefore:

$$A = 27'0'' = \times\ 37'0'' = 999.00 \text{ square feet}$$

$$H = \frac{1{,}475 \text{ cubic feet}}{999.0 \text{ square feet}} = 1.48 \text{ feet, or } 1'6''$$

A dike wall height of 1′6″ will contain the volume of the largest tank plus ten percent. As previously stated, the tanks are raised two inches above the dike floor. Therefore, sixteen inches of tank capacity for the two tanks other than the largest tank is below the top of the dike. This volume must also be contained.

A.9 Unavailable Capacity. Through use of a tank capacity conversion table, we determine that this volume in each tank below the top of the dike is 663 gallons. This means that we must provide for 1,326 gallons additional capacity, or 177.3 cubic feet.

$$H = \frac{663 \text{ gallons} \times 2 \text{ tanks}}{7.48 \text{ gallons per cubic feet}} = 177.3 \text{ cubic feet}$$

$$H = \frac{(1{,}475.4 \text{ cubic feet} + 177.3 \text{ cubic feet})}{(999.0 \text{ square feet})}$$

$$H = 1.65 \text{ feet, or } 1'8''$$

Dimensions for containment installation:

Length = 37′0″
Width = 27′0″
Height = At least 1′8″ (rounded to the next higher full inch)

A.10 Additional Requirements. Dike walls may be sloped, complicating the calculation somewhat. Higher dikes require provisions for entry and firefighting. Distances from dikes to property lines and buildings can significantly affect the layout.

APPENDIX G

Tank Data Sheet

This appendix contains a tank data sheet that should be completed by the purchaser when specifiying a field-erected AST. When filled out completely, it will provide the manufacturer with adequate information to design and construct an AST and compare proposals from the various manufacturers more effectively.

Aboveground Steel Storage Tank Handbook—Tank Data Sheet

A. User/Owner Information

User/Owner ______________________________
Name of plant ______________________________
Address ______________________________
City ______________________ State ____ Zip code ______-___
Phone ______________________ Fax ______________________

Purchaser ______________________________
Address ______________________________
City ______________________ State ____ Zip code ______-___
Phone ______________________ Fax ______________________

B. Tank Description

Tank no(s). ______________ Quantity _______ Tank diameter ________ feet
Tank height: ☐ Specific tank height __________ feet
☐ Nominal capacity ____________ (gallons, barrels, cubic feet)
☐ Working capacity[1] ____________ (gallons, barrels, cubic feet)
Foundation: ☐ Earth ☐ Concrete ringwall ☐ Other ____________
Foundation constructed by: ☐ Tank erector ☐ User/Owner

C. Product Information

Product description ______________ Specific gravity ______ At ______ °F
Vapor pressure __________ psia Flash point __________ °F

D. Design Information

Design specific gravity __________ Design metal temperature[2] ________ °F
Maximum operating temperature __________ °F
Design over-pressure: ☐ Atmospheric ☐ Other _________ (psig, inches of water)
Design vacuum load: ☐ None ☐ Other _________ (psig, inches of water)
Corrosion allowance: Shell _________ inches Roof _________ inches
Bottom _________ inches Structural ________ inches
Secondary bottom _______ inches
Design standard: ☐ API 650
Appendix A design allowed ☐ Yes ☐ No
Joint efficiency: ☐ 0.70 ☐ 0.85
☐ Appendix J design, shop-assembled tank
☐ API 620
Joint efficiency[3]: ☐ 0.85 ☐ 1.0
☐ AWWA D-100
Appendix C design allowed ☐ Yes ☐ No
☐ Other ______________

Tank Data Sheet (*continued*)

E. Tank Roof Design:

Design Loads: Uniform live load ☐ 25 psf ☐ Other ____________ psf
Snow load ☐ None ☐ Other ____________ psf
Frangible joint check[4] ☐ Required ☐ Not required

F. Tank Construction

Bottom: Slope: ________ inches per foot ☐ Crowned ☐ Coned
Primary bottom nominal[5] thickness ________ inches
Weld type: ☐ Single lap ☐ Single butt weld ☐ Double butt weld
Release prevention barrier required ☐ Yes ☐ No
☐ Secondary steel plate
Secondary bottom nominal thickness ________ inches
Type of spacer between bottoms: ________________________
☐ Flexible liner —type: ________________________
☐ Other: ________________________

Shell: Minimum nominal thickness: ☐ Per design standard ☐ ______ inches
Minimum corroded thickness: ☐ Per design standard ☐ ______ inches

Roof: ☐ Fixed roof[6]
☐ Self-supported cone roof — slope: ______ inches per foot
☐ Supported cone roof — slope: ☐ 3/4:12 ☐ Other _______ :12
☐ Self-supported umbrella roof[7] — radius: _________ feet
☐ Self-supported dome roof — radius: _________ feet
☐ Floating roof[8]
☐ Internal floating roof — type ________________________
☐ External floating roof — type: ________________________
☐ Other ________________________

G. Tank Loadings

Wind Design: Wind load velocity _____________ mph
Shell wind stability per ☐ API 650 ☐ Other _____________
Intermediate wind girders allowed ☐ Yes ☐ No
Top wind girder used as a walkway ☐ Yes ☐ No

Earthquake Design: ☐ Required ☐ Not required
☐ API 650 Appendix E design ☐ Other _____________
Seismic zone ☐ Figure E-1 ☐ Specified __________
Importance factor (I): ☐ 1.0 ☐ Other ______
Site coefficient (S): ☐ Unknown (1.5) ☐ Other ______
Tie rods required (API 650 3.10.4.5) ☐ Yes ☐ No

Rainfall: Maximum total rainfall __________ inches
Maximum rainfall rate ___________ inches/hour

Tank Data Sheet (*continued*)

H. Testing

Shell: Radiograph examination required: ☐ Yes ☐ No
Test weld seams by: ☐ Hydrotest ☐ Vacuum box ☐ Other __________
Bottom: Test weld seams by: ☐ Vacuum box ☐ Other __________
Roof: Weld testing: ☐ None ☐ Vacuum box ☐ Pressure test ☐ Other __________

I. Venting Design

Designed by: ☐ User/Owner
Nominal nozzle size __________ inch
Vent type: __________
Vent supplied by: ☐ Owner ☐ Tank erector
☐ Tank erector
Pumping rates:
In __________ (bph, gpm) Out __________ (bph, gpm)
☐ Process engineer

J. Appurtenances

Mark	Quantity	Size	Flange type	Location[9]	Description

[1] Provide a sketch to define the working capacity as discussed in Chapter 13 and depicted in Figure 13-1.
[2] Definition of the design metal temperature is given in the shell design section of Chapter 9.
[3] See the shell design section in Chapter 9 on a difference between the API 620 joint efficiencies.
[4] Discussion of frangible joint is given in the roof design section of Chapter 9 and Chapter 13.
[5] Nominal thickness includes any corrosion allowance.
[6] Discussion on the different fixed roofs options is discussed in the roof design section of Chapter 9 and Chapter 13.
[7] See Figure 9-1(c) on the difference between umbrella and dome roofs.
[8] Discussion on the types of floating roofs is given in the floating roof design section of Chapter 9 and Chapter 13.
[9] Specify if the nozzle is located in the shell or roof of the tank.

Glossary of Industry Acronyms and Abbreviations

ACP: area contingency plan
ACT: alternate control techniques
AE: acoustic emissions
AFV: alternative fuel vehicle
ANSI: American National Standards Institute
API: American Petroleum Institute (Washington, DC)
ASME: American Society of Mechanical Engineers
AST: aboveground storage tank
ASTM: American Society for Testing and Materials (Philadelphia, PA)
AWWA: American Water Works Association

BAAQMD: Bay Area Air Quality Management District, California
BACM: best available control measures, under CAAA
BACT: best available control technology, under CAA
BAT: best available technology
BCT: best conventional technology
BDAT: best demonstrated available technology
BIFs: boilers and industrial furnaces, RCRA
BOCA: Building Officials and Code Administrators
BOD: biological oxygen demand

b/d: barrels per day
BTEX: benzene, toluene, ethyl benzene, and xylene
Btu: British thermal unit

CAA: Clean Air Act; passed in 1970, reauthorized in 1977 and 1990
CAAA: Clean Air Act Amendments of 1990
CCME: Canadian Council of Ministers for the Environment; drafts ECPs in Canada
CEM: continuous emissions monitoring
CERCLA: Comprehensive Environmental Response, Compensation and Liability Act (also known as Superfund)
CFC: chlorofluorocarbon
CFR: Code of Federal Regulations
CGL: comprehensive general liability insurance
CO: carbon monoxide
CO_2: carbon dioxide
COD: chemical oxygen demand
CTGs: control technique guidelines
CWA: Clean Water Act; passed in 1972, updated in 1987

DFT: day film thickness
DMT: design metal temperature
DNAPLs: dense non-aqueous-phase liquids
DOE: Department of Energy, U.S.
DOT: Department of Transportation, U.S.

ECP: Environmental Code of Practice, as drafted by CCME
EDF: Environmental Defense Fund (Washington, DC)
EFR: external floating roof
EHS: extremely hazardous substance
EP: extraction procedure, RCRA
EPA: U.S. Environmental Protection Agency
EPCRA: Emergency Planning and Community Right-to-Know Act; see SARA Title III
ERNS: Emergency Response Notification System, under EPA
ET: environmental response team

FR: Federal Register
FRP: facility response plan
FWPCA: Federal Water Pollution Control Act

GACT: generally available control technology

GPM: gallons per minute

HAP: hazardous air pollutant, under CAAA
HCFC: hydrochlorofluorocarbon
HCS: hazard communication standard
HON: hazardous organic NESHAP, MACT standards for chemical manufacturing industry
HSWA: Hazardous and Solid Waste Amendment

IFCI: International Fire Codes Institute
IFR: internal floating roof
ILTA: Independent Liquid Terminals Association (Washington, DC)

LAER: lowest achievable emissions rate
LEPC: local emergency planning committee
LUST: leaking underground storage tank

MACT: maximum achievable control technology, under CAAA
MDMT: minimum design metal temperature
MGD: million gallons/day
MSDS: material safety data sheet
MTBE: methyl tertiary butyl ether

NAAQS: National Ambient Air Quality Standards, under CAA
NACE: National Association of Corrosion Engineers
NCP: National Contingency Plan
NESHAP: National Emission Standards for Hazardous Air Pollutants, under CAA
NFC: National Fire Code, Canada
NFPA: National Fire Protection Association (Quincy, MA)
NO_x: nitrogen oxides
NO_2: nitrogen diozide
NOI: notice of intent
NPDES: National Pollutant Discharge Elimination System
NPL: National Priorities List, under CERCLA
NPRA: National Petroleum Refiners Association (Washington, DC)
NRC: National Response Center
NRT: National Response Team
NSPS: New Source Performance Standards

O_2: oxygen

O_3: ozone
ODS: ozone-depleting substance
OMB: Office of Management and Budget
OPA: Oil Pollution Act of 1990
OSC: on-scene coordinator
OSHA: Occupational Safety and Health Administration
OWEC: EPA Office of Wastewater Enforcement and Compliance

Pb: lead
PCBs: polychlorinated biphenyls
PCP: pentachlorophenol
PEI: Petroleum Equipment Institute (Tulsa, OK)
PELs: permissible exposure limits, OSHA
PMAA: Petroleum Marketers Association of America (Washington, DC)
POTW: publicly owned treatment works
PPA: Pollution Prevention Act of 1990
ppb: parts per billion
ppg: pounds per gallon
ppm: parts per million
PRP: potentially responsible party, under CERCLA
PSD: prevention of significant deterioration, under CAA
psia: pounds per square inch atmospheric
psig: pounds per square inch gauge

RA: regional administrator
RACM: reasonably available control measures, under CAAA
RACT: reasonably available control technology, under CAAA
RCRA: Resource Conservation and Recovery Act (1976, 1984)
RCT: reference control technology
RFG: reformulated gasoline
RFP: reasonable further progress, under CAA
RPB: release prevention barrier
RRT: regional response team

SARA: Superfund Amendments and Reauthorization Act of 1986, expanded CERCLA
SARA Title III: section of SARA requiring public disclosure of chemical release information
SBCCI: Southern Building Code Congress International

SCAQMD: South Coast Air Quality Management District, California
SERC: state emergency response commission, RCRA
SFPC: Standard Fire Prevention Code, SBCCI
SIGMA: Society of Independent Gasoline Marketers
SIC: standard industrial classification
SIP: state implementation plan, under CAA
SO_2: sulfur dioxide
SOCMI: Synthetic Organic Chemical Manufacturers Industry
SPCC: spill prevention, control, and countermeasures plan
SSPC: Steel Structures Painting Council
STA: Steel Tank Association (Lake Zurich, IL)
STAC: Steel Tank Association Canada
STI: Steel Tank Institute (Lake Zurich, IL)

TC: toxicity characteristics
TCLP: toxicity characteristics leaching procedure, under RCRA
TPQ: threshold planning quantity, RCRA
TPY: tons per year
TRI: Toxic Release Inventory, under SARA Title III
TSCA: Toxic Substances Control Act
TSDF: treatment, storage, or disposal facility, under RCRA
TS: total solids
TSS: total suspended solids

UBC: Uniform Building Code
UFC: Uniform Fire Code
UL: Underwriters Laboratories
ULC: Underwriters Laboratories, Canada
USC: United States Code
USCG: United States Coast Guard
UST: underground storage tank
UV: ultraviolet

VOC: volatile organic compound
VOL: volatile organic liquid
VS: volatile solids
VSS: volatile suspended solids

WQA: Water Quality Act

Index

Printed and bound by CPI Group (UK) Ltd, Croydon, CR0 4YY

16/09/2026

14963462-0001